**Progress in Probability and Statistics**
**Vol. 9**

**Edited by**
**Peter Huber**
**Murray Rosenblatt**

Birkhäuser
Boston · Basel · Stuttgart

# Seminar on Stochastic Processes, 1984

E. Çınlar
K. L. Chung
R. K. Getoor
editors

1986

Birkhäuser
Boston · Basel · Stuttgart

Editors:

E. Çınlar
Civil Engineering Department
Princeton University
Princeton, New Jersey 08544

R. K. Getoor
Department of Mathematics
University of California, San Diego
La Jolla, California 92093

K. L. Chung
Department of Mathematics
Stanford University
Stanford, California 94305

**Library of Congress Cataloging in Publication Data**

Seminar on Stochastic Processes (4th : 1984 :
Northwestern University)
Seminar on Stochastic Processes, 1984.

(Progress in probability and statistics ; vol. 9)
Papers presented during the fourth seminar, held at
Northwestern University, Evanston.
1. Stochastic processes -- Congresses. I. Çınlar, E.
(Erhan), 1941– . II. Chung, Kai Lai, 1917–
III. Getoor, R. K. (Ronald Kay), 1929– .
IV. Title. V. Series: Progress in probability
and statistics ; v. 9.
QA274.A1S45 1984 519.2 85–22961
ISBN-13: 978-1-4684-6747-5

**CIP-Kurztitelaufnahme der Deutschen Bibliothek**

**Seminar on Stochastic Processes:**
Seminar on Stochastic Processes . . . – Boston ; Basel ;
Stuttgart : Birkhäuser 1986
(Progress in probability and statistics ; Vol. 9)
ISBN-13: 978-1-4684-6747-5

NE: GT


Softcover reprint of the hardcover 1st edition 1986
ISBN-13: 978-1-4684-6747-5 e-ISBN-13: 978-1-4684-6745-1
DOI: 10.1007/978-1-4684-6745-1

## TABLE OF CONTENTS

## FOREWORD

This volume consists of about half of the papers presented during a three-day seminar on stochastic processes held at Northwestern University, Evanston. The seminar was the fourth of such yearly seminars aimed at bringing together a small group of researchers to discuss their current work in an informal atmosphere.

The invited participants in the seminar were B.W. ATKINSON, R.M. BLUMENTHAL, K. BURDZY, D. BURKHOLDER, M. CRANSTON, C. DOLEANS-DADE, J.L. DOOB, N. FALKNER, P. FITZSIMMONS, J. GLOVER, F. KNIGHT, T. McCONNELL, J.B. MITRO, S. OREY, J. PITMAN, A.O. PITTENGER, Z. POP-STOJANOVIC, P. PROTTER, T. SALISBURY, M. SHARPE, C.T. SHIH, A. SZNITMAN, S.J. TAYLOR, J. WALSH, and R. WILLIAMS. We thank them and the other participants for the lively seminar they created.

The seminar was made possible through the partial support of the Air Force Office of Scientific Research via their Grant No. 82-0189 to Northwestern University.

E.Ç.
Princeton, 1985

*Seminar on Stochastic Processes, 1984*
Birkhäuser, Boston, 1986

# TWO-SIDED TIME-HOMOGENEOUS MARKOV PROCESSES

by

Bruce W. Atkinson

## 1. Introduction

The purpose of this paper is to discuss conditions under which it is possible to construct a Markov process indexed by $\mathbb{R}$ with random times of birth and death and which is time-homogeneous in both directions.

In section 2 we review a few facts concerning Markov processes indexed by $\mathbb{R}_+$, and we introduce some terminology useful for the two-sided construction. Implicit in the statement that a Markov process indexed by $\mathbb{R}_+$ is time-homogeneous is a collection $(P_t : t \in \mathbb{R}_+)$ of sub-Markov kernels on the state space $(E, \mathcal{E})$ which serve as transitions. The collection $(P_t)$ need only satisfy the Chapman-Kolmogorov equations in the "loose sense", i.e. $\forall\, f \in \mathcal{E}^+$, $s,t \in \mathbb{R}_+$ we have $P_{s+t}f(x) = P_s(P_tf)(x)$ for a.a. x relative to any fixed one-dimensional distribution of the process. We call the family $(P_t)$ *representable* (see(2.7)) if $\forall\, x \in E$ there exists a time-homogeneous Markov process $P^x$ (i.e. a certain type of measure on path space) so that $P^x$ is a probability, $P^x(X_0 = x) = 1$, and $P^x$ has transitions $(P_t)$. (Here, for $t \in \mathbb{R}_+$, $X_t$ is the t-th coordinate on path space; see section 2 for all the precise definitions.) If the family is representable then it is a semi-group and the family

$(P^x)$ is *self-subordinate* in the sense that $\forall\, x \in E$, positive measurable $Y$, and $t \in \mathbb{R}_+$ it follows that $P^x[(Y\circ\theta_t)|X_t] = P^{X_t}Y$; see (2.10). In fact the property of being self-subordinate characterizes all families $(P^x)$ of probabilities on path space which satisfy $P^x(X_0 = x) = 1$ $\forall\, x \in E$ and which arise from a semi-group $(P_t)$ as described above.

In section 3 the concepts introduced in the preceding paragraph are extended to the two-sided case. As in the one-sided case, implicit in the definition of time-homogeneity in both directions are families $(P_t : t \in \mathbb{R}_+)$ and $(Q_t : t \in \mathbb{R}_+)$ of sub-Markov kernels which serve as forward and backward transitions. As before, $(P_t)$ and $(Q_t)$ need only satisfy the Chapman-Kolmogorov equations in the loose sense; see (3.2), (3.4). However, if both $(P_t)$ and $(Q_t)$ are representable, as described above, then there is a family of probabilities $(P^x : x \in E)$ on two-sided path space so that $\forall\, x \in E$, $P^x(X_0 = x) = 1$, and every two-sided time-homogeneous Markov process $P$ (i.e. a certain type of measure on two-sided path space; see section 3 for details) with forward and backward transitions $(P_t)$ and $(Q_t)$ is *subordinate* to $(P^x)$ in the sense that for every positive measurable $Y$, and $\forall\, t \in \mathbb{R}$ we have $P(Y\circ\theta_t|X_t) = P^{X_t}Y$, where $X_t$ is the t-th coordinate in path space and $\theta_t$ is the shift operator. However, rarely will it happen that each $P^x$ is itself a time-homogeneous Markov process with transitions $(P_t)$ and $(Q_t)$. In fact this can only happen in a deterministic case; see (3.16). This differs from the one-sided case. Loosely speaking, the difference lies in the following remark: Whereas in the one-sided case to construct a process one usually only needs a semigroup $(P_t)$ and an arbitrary measure to serve as the distribution of the initial state $X_0$, in the two-sided case one needs that the forward and backward transitions $(P_t)$ and $(Q_t)$ be knit together with the one-dimensional distributions of the process, $(\pi_t : t \in \mathbb{R})$, in an obvious way; see (3.5). There does not seem to be nearly as much freedom in selecting the distribution of a two-sided process at a fixed time, say $t = 0$.

Section 4, in response to the limitations described above, gives a partial answer, in the form of a sufficient condition, to the question: Given a family of sub-Markov kernels $(P_t)$ and a measure $\pi$ on the state space, when can one construct a two-sided time-homogeneous Markov process with forward transitions $(P_t)$ and with $\pi$ as the distribution at $t = 0$? The sufficient condition is basically that there exist $\lambda \in \mathbb{R}$ so that $\forall\, t \in \mathbb{R}_+$, $\pi P_t \leq e^{\lambda t}\pi$; see (4.3). The process P constructed in (4.3) is called *quasi-stationary* since $\forall\, t \in \mathbb{R}$, $P \circ \theta_t = e^{\lambda t} P$. (The case $\lambda = 0$ corresponds with the theory of weak duality, and the associated two-sided process is stationary.) An equality of the form $\pi P_t = e^{\lambda t}\pi$ $\forall\, t \in \mathbb{R}_+$ plays a role in the study of Brownian motions on manifolds; see [5] for related discussion. Perhaps quasi-stationary processes will be useful in the study of $\lambda$-potential theory; again see [5].

The same issues, as introduced above, were discussed in [1] for discrete time and space. As one might expect, more complete results are possible. In fact a complete characterization is given in [1] for two-sided time-homogeneous Markov chains with an irreducible forward one-step transition matrix which has a power $\geq 1$ with nonzero diagonal entries. A special case of this characterization led to the definition of *quasi-stationary chains*. Quasi-stationary processes in this paper are the appropriate analogs of quasi-stationary chains.

## 2. One-Sided Time-Homogeneous Markov Processes

In this section we shall discuss certain aspects of the usual one-sided Markov process theory for two reasons: to set up notation used later, and to motivate definitions in section 3 for the two-sided case.

Let $(E,\mathcal{E})$ be a measurable space. Throughout this paper we assume that $\forall\, x \in E$, $\{x\} \in \mathcal{E}$. Let c be a fixed element not in E.

(2.1) DEFINITION. (a) $\Omega_+^c = \{$functions $\omega: \mathbb{R}_+ \to E \cup \{c\}$: $\omega(0) \in E$, $\omega(s) = c \Rightarrow \omega(t) = c \;\; \forall\, t > s$, and $\{t: \omega(t) \in E\}$ is open relative to $\mathbb{R}_+\}$.

(b) *Define*, $\forall\, t \in \mathbb{R}_+$, $X_t: \Omega_+^c \to E \cup \{c\}$ by $X_t(\omega) = \omega(t)$.

(c) *Define*, $\forall\, t \in \mathbb{R}_+$, $\theta_t: \Omega_+^c \to \Omega_+^c$ by $[\theta_t(\omega)](s) = \omega(s+t)$.

(d) $\mathcal{F}_+^c = \sigma(X_t: t \in \mathbb{R}_+)$.

(2.2) REMARK. If $f \in \mathcal{E}^+$, then we make the convention that f is extended to $E \cup \{c\}$ by setting $f(c) = 0$. Also, in (2.1d) it is implicit that $E \cup \{c\}$ is equipped with the σ-field generated by $\mathcal{E}$ and $\{c\}$.

We now set some conventions in this paper for ease of exposition. Let $(B,\mathcal{B},\mu)$ be a measure space. We shall use the same symbol for a measure and its integrals. Thus, if $f \in \mathcal{B}^+$ then $\mu(f)$, or sometimes simply $\mu f$, shall stand for $\int \mu(dx)f(x)$. Also, suppose $\mathcal{C}$ is a sub-σ-field of $\mathcal{B}$ and $(B,\mathcal{C},\mu)$ is σ-finite. Then for $f \in \mathcal{B}^+$ we shall denote by $\mu(f|\mathcal{C})$ the $\mathcal{C}$-measurable $[0,\infty]$-valued variable (determined only μ a.e.) so that $\forall\, g \in \mathcal{C}^+$, $\mu(fg) = \mu(\mu(f|\mathcal{C})g)$. (Of sourse the existence of $\mu(f|\mathcal{C})$ is a consequence of the Radon-Nikodym theorem.)

(2.3) DEFINITION. *Let* P *be a measure on* $(\Omega_+^c, \mathcal{F}_+^c)$ *and suppose that,* $\forall\, t \in \mathbb{R}_+$, *there exists a σ-finite measure* $\pi_t$ *on* $(E,\mathcal{E})$ *so that* $Pf(X_t) = \pi_t f$, $\forall\, f \in \mathcal{E}^+$. *Then* P *is called a (one-sided)* Markov process *if*

$$P(f(X_t)|\sigma(X_u: u \leq s)) = P(f(X_t)|\sigma(X_s)) \text{ on } \{X_s \in E\}$$

whenever $0 \leq s \leq t < \infty$ and $f \in \mathcal{E}^+$.

Recall that a sub-Markov kernel on $(E,\mathcal{E})$ (or sometimes written "from $(E,\mathcal{E})$ to $(E,\mathcal{E})$") is a function $K\colon E \times \mathcal{E} \to [0,1]$ satisfying

(i) $\forall\, x \in E$, $A \to K(x,A)$ is a measure, and

(ii) $\forall\, A \in \mathcal{E}$, $x \to K(x,A)$ is $\mathcal{E}$-measurable.

Also, for $f \in \mathcal{E}^+$ we define $Kf \in \mathcal{E}^+$ by $Kf(x) = \int K(x,dy)f(y)$.

(2.4) DEFINITION. *Let* P *be Markov as in* (2.3). P *is called* time-homogeneous *if* $\forall\, t \in \mathbb{R}_+$ *there exists a sub-Markov kernel* $P_t$ *on* $(E,\mathcal{E})$ *with the following property:* $\forall\, s \in \mathbb{R}_+$, *and* $\forall\, f,g \in \mathcal{E}^+$, *we have* $P(f(X_s)g(X_{s+t})) = \pi_s(f(P_tg))$. *In this case we say that* P *has* transitions $(P_t)$.

(2.5) PROPOSITION. *Let* P *be a time-homogeneous Markov process with transitions* $(P_t)$.

(a) $\forall\, r,s,t \in \mathbb{R}_+$, *and* $\forall\, f \in \mathcal{E}^+$, $P_{s+t}f(x) = P_s(P_tf)(x)$ *for* $\pi_r$ a.e. x.

(b) *Let* $r \in \mathbb{R}_+$, *and* $(t_n)$ *a decreasing sequence of real numbers in* $\mathbb{R}_+$ *with* $t_n \to 0$ as $n \to \infty$. *Then* $P_{t_1}(x,E) \leq P_{t_2}(x,E) \leq \cdots \to 1$ *for* $\pi_r$ a.e. x.

PROOF. (a) Follows easily from the definitions.

(b) By the definition of $\Omega_+^c$, $1_E(X_{r+t_n}) \leq 1_E(X_{r+t_{n+1}})$. Hence $\forall g \in \mathcal{E}^+$,

$$P(g(X_r)1_E(X_{r+t_n})) \leq P(g(X_r)1_E(X_{r+t_{n+1}}))$$

or

$$\int \pi_r(dx)g(x)P_{t_n}(x,E) \leq \int \pi_r(dx)g(x)\, P_{t_{n+1}}(x,E).$$

Hence $P_{t_n}(x,E) \leq P_{t_{n+1}}(x,E)$ for $\pi_r$ a.e. x. Also on $\{X_r \in E\}$, $1_E(X_{r+t_n}) \uparrow$

1 as $n \to \infty$ by the definition of $\Omega_+^c$. Thus $\forall g \in \mathcal{E}^+$,

$$\int \pi_r(dx)g(x)P_{t_n}(x,E) \uparrow \int \pi_r(dx)g(x)$$

as $n \to \infty$, and the desired result follows. □

(2.6) REMARK. The property (2.5a) is a slight extension of the concept, found in [2], of a transition function in the loose sense.

(2.7) DEFINITION. *A family* $(P_t: t \in \mathbb{R}_+)$ *of sub-Markov kernels on* $(E,\mathcal{E})$ *is called* representable *if* $\forall x \in E$ *there exists a time-homogeneous Markov process* $P^x$ *with transitions* $(P_t)$ *so that* $P^x$ *is a probability and* $P^x(X_0 = x) = 1$. *In this case we say that* $(P_t)$ *is represented by the family* $(P^x)$, *or that* $(P^x)$ *represents* $(P_t)$.

The next result follows from definition and (2.5).

(2.8) PROPOSITION. *Let* $(P_t)$ *be a representable family.*

(a) $\forall s,t \in \mathbb{R}_+$ *and* $\forall f \in \mathcal{E}^+$, $P_{s+t}f = P_s(P_tf)$.

(b) $\forall x \in E$ *and* $\forall f \in \mathcal{E}^+$ *and* $\forall t \in \mathbb{R}_+$, $P_tf(x) = P^xf(X_t)$.

(c) $\forall x \in E$, $P_t(x,E) \uparrow 1$ *as* $t \downarrow 0$.

(d) $\forall Y \in (F_+^c)^+$, $x \to P^xY$ is $\mathcal{E}^+$*-measurable.*

(2.9) REMARKS. (a) (2.8a) means that $(P_t)$ is a transition function in the usual sense (i.e. it is a semigroup).

(b) (2.8b) says that we can recover $(P_t)$ from $(P^x)$, and thus $(P^x)$ would be more interesting to study.

(2.10) DEFINITION. *Let* $(Q^x: x \in E)$ *be a family of probabilities on*

$(\Omega_+^c, F_+^c)$ *so that* $\forall\, x \in E$, $Q^x(X_0 = x) = 1$, *and* $\forall\, Y \in (F_+^c)^+$ $x \to Q^x Y$ *is* $E$*-measurable. Next, let* $Q$ *be a measure on* $(\Omega_+^c, F_+^c)$ *so that* $\forall\, t \in \mathbb{R}_+$ *there exists a* $\sigma$*-finite measure* $\mu_t$ *on* $(E,E)$ *with* $Qf(X_t) = \mu_t f \quad \forall\, f \in E^+$. *Then we say that* $Q$ *is* subordinate to the family $(Q^x)$ *if* $\forall\, t \in \mathbb{R}_+$ *and* $Y \in (F_+^c)^+$,

$$Q(Y\circ\theta_t | \sigma(X_t)) = Q^{X_t} Y \text{ on } \{X_t \in E\}.$$

(2.11) PROPOSITION. *Let* $P$ *be a time-homogeneous Markov process with transitions* $(P_t)$. *Suppose that* $(P_t)$ *is represented by the family* $(P^x)$. *Then* $P$ *is subordinate to* $(P^x)$.

PROOF. By the monotone class theorem it suffices to check that $P(g(X_t)Y\circ\theta_t) = P(g(X_t)P^{X_t}Y)$ where $t \in \mathbb{R}_+$, $g \in E^+$ with $Pg(X_t) < \infty$, and $Y$ has the form $Y = f_1(X_{t_1})\cdots f_n(X_{t_n})$ with $f_1,\ldots,f_n \in E^+$ and $0 \leq t_1 < t_2 < \cdots < t_n < \infty$. But by the fact that $P$ has transitions $(P_t)$,

$$P(g(X_t)\ Y\circ\theta_t) = P(g(X_t)\ f_1(X_{t_1+t})\ \cdots\ f_n(X_{t_n+t}))$$

$$= P(g(X_t)\ f_1(X_{t_1+t})\ \cdots\ f_{n-1}(X_{t_{n-1}+t})\ P_{t_n-t_{n-1}}f_n(X_{t_{n-1}}))$$

$$= \cdots = P(g(X_t)[P_{t_1}(f_1P_{t_2-t_1}(\cdots(f_{n-1}P_{t_n-t_{n-1}}f_n)\cdots))(X_t)]).$$

But since $P^x$ is Markov with transitions $(P_t)$,

$$P^xY = P^x(f_1(X_{t_1})\cdots f_n(X_{t_n}))$$

$$= P^x(f_1(X_{t_1})\cdots f_{n-1}(X_{t_{n-1}})\ P_{t_n-t_{n-1}}f_n(X_{t_{n-1}}))$$

$$= \cdots = P^x(f_1(X_{t_1})[P_{t_2-t_1}(\cdots(f_{n-1}P_{t_n-t_{n-1}}f_n)\cdots)(X_{t_1})])$$

$$= P_{t_1}(f_1P_{t_2-t_1}(\cdots(f_{n-1}P_{t_n-t_{n-1}}f_n)\cdots))(x),$$

which is what was to be shown.

(2.12) REMARK. Let $(P_t)$ be represented by $(P^x)$. Then since each $P^x$ is time-homogeneous with transitions $(P_t)$, (2.11) implies that, $\forall\, y \in E$, $P^y$ is subordinate to $(P^x)$. That is, the family $(P^x)$ is *self-subordinate.* In fact the next result states that self-subordination characterizes the families of probabilities which represent, in the sense of (2.7), a family of sub-Markov kernels.

(2.13) THEOREM. *Let* $(Q^x : x \in E)$ *be a family of probabilities on* $(\Omega_+^c, F_+^c)$ *so that* $\forall\, x \in E$, $Q^x(X_0 = x) = 1$, *and* $\forall\, Y \in (F_+^c)^+$, $x \to Q^x Y$ *is E-measurable. Further suppose that,* $\forall\, y \in E$, $Q^y$ *is subordinate to* $(Q^x)$ *(i.e. the family* $(Q^x)$ *is self-subordinate). For* $t \in \mathbb{R}_+$ *and* $A \in E$ *define* $Q_t(x,A) = Q^x(X_t \in A)$ $\forall\, x \in E$. *Then* $(Q^x)$ *represents* $(Q_t)$.

PROOF. We will use induction at first to prove the statement: If $n \geq 1$, $0 < t_1 < \cdots < t_n < \infty$, and $f_1,\cdots,f_n \in E^+$, then

$$Q^x(f_1(X_{t_1})\cdots f_n(X_{t_n})) = Q_{t_1}(f_1Q_{t_2-t_1}(\cdots(f_{n-1}Q_{t_n-t_{n-1}}f_n)\cdots))(x).$$

*(n=1)*: $Q^x f_1(X_{t_1}) = Q_{t_1} f_1(x)$ by definition.

*(induction step)*: Suppose our statement is true for n. Let $0 < t_1 < \cdots < t_n < t_{n+1} < \infty$ and $f_1,\cdots,f_n,f_{n+1} \in E^+$. Then

$$Q^x(f_1(X_{t_1})\cdots f_{n+1}(X_{t_{n+1}}))$$

$$= Q^x(f_1(X_{t_1})[(f_2(X_{t_2-t_1})\cdots f_{n+1}(X_{t_{n+1}-t_1}))\circ\theta_{t_1}])$$

$$= Q^x(f_1(X_{t_1})h(X_{t_1})),$$

where

$$h(y) = Q_{t_2-t_1}(f_2 Q_{t_3-t_2}(\cdots(f_n Q_{t_{n+1}-t_n} f_{n+1})\cdots))(y).$$

But since

$$Q^x(f_1(X_{t_1})h(X_{t_1})) = Q_{t_1}(f_1 h)(x),$$

our statement holds for n+1. Thus the statement holds for all $n \geq 1$ by induction.

It is now obvious, by a monotone class argument, that if $0 < s < t$ and $f \in \mathcal{E}^+$, then $Q^x(f(X_t)|\sigma(X_u: u \leq s)) = Q_{t-s}f(X_s)$ on $\{X_s \in E\}$, and thus each $Q^x$ is time-homogeneous Markov, which is the desired conclusion. □

## 3. Two-Sided Time-Homogeneous Markov Processes

Let $(E,\mathcal{E})$ be as in section 2 and a an element not in E. Next, let b be an element not in $E \cup \{a\}$. Whenever it comes into play, the σ-field on $E \cup \{a,b\}$ is that generated by $\mathcal{E}$, $\{a\}$, and $\{b\}$. Also, we automatically extend any $f \in \mathcal{E}^+$ to $E \cup \{a,b\}$ by setting $f(a) = f(b) = 0$.

(3.1) DEFINITION. (a) $\Omega = \{$functions $\omega: \mathbb{R} \to E \cup \{a,b\}$: $\omega(t) \in E$ for some t, $\omega(t) = a \Rightarrow \omega(s) = a$ for $s < t$, $\omega(t) = b \Rightarrow \omega(s) = b$ for $s > t$, and $\{t: \omega(t) \in E\}$ is open$\}$.

(b) $\forall\, t \in \mathbb{R}$, $X_t: \Omega \to E \cup \{a,b\}$ is defined by $X_t(\omega) = \omega(t)$.

(c) $\forall\, t \in \mathbb{R}$, $\theta_t\colon \Omega \to \Omega$ is defined by $[\theta_t(\omega)](s) = \omega(s+t)$.

(d) $\mathcal{F} = \sigma(X_t\colon t \in \mathbb{R})$.

(3.2) DEFINITION. (a) *Let* P *be a measure on* $(\Omega,\mathcal{F})$ *and suppose that* $\forall\, t \in \mathbb{R}$ *there exists a* $\sigma$*-finite measure* $\pi_t$ *on* $(E,\mathcal{E})$ *so that* $Pf(X_t) = \pi_t f \quad \forall\, f \in \mathcal{E}^+$. *Then,* P *is called a* two-sided Markov process *if*

$$P(f(X_t)\,|\,\sigma(X_u\colon u \leq s)) = P(f(X_t)\,|\,\sigma(X_s)) \text{ on } \{X_s \in E\}$$

whenever $s \leq t$ and $f \in \mathcal{E}^+$.

(b) *Let* P *be Markov as in* (a). *Then* P *is called a* two-sided time-homogeneous Markov process *if there exist families of sub-Markov kernels on* $(E,\mathcal{E})$, $(P_t\colon t \in \mathbb{R}_+)$ *and* $(Q_t\colon t \in \mathbb{R}_+)$, *so that we have*

$$P(f(X_s)g(X_t)) = \pi_s(f(P_{t-s}g)) = \pi_t(g(Q_{t-s}f))$$

whenever $-\infty < s < t < \infty$, and $f,g \in \mathcal{E}^+$.

(3.3) REMARK. It is easy to check that if P is a measure with $P(X_t \in dx)$ $\sigma$-finite on $(E,\mathcal{E})$ $\forall\, t$, then the condition in (3.2a) is equivalent to: $P(f(X_s)\,|\,\sigma(X_u\colon u \geq t)) = P(f(X_s)\,|\,\sigma(X_t))$ on $\{X_s \in E\}$ whenever $s \leq t$ and $f \in \mathcal{E}^+$. Thus if P, $(P_t)$, and $(Q_t)$ are as in (3.2b), then if $s \leq t$ and $f \in \mathcal{E}^+$ we have that $P(f(X_t)\,|\,(X_u\colon u \leq s)) = P_{t-s}f(X_s)$ on $\{X_s \in E\}$ and $P(f(X_s)\,|\,(X_u\colon u \geq t)) = Q_{t-s}f(X_t)$ on $\{X_t \in E\}$. Thus, in this case, from now on we shall say that P is two-sided time-homogeneous Markov with *forward and backward transitions* $(P_t)$ *and* $(Q_t)$.

The next result follows in the same manner as in the proof of (2.5), and the proof is thus omitted.

(3.4) PROPOSITION. *Let* P *be a two-sided time-homogeneous Markov process with forward and backward transitions* $(P_t)$ *and* $(Q_t)$.

(a) $\forall\ r \in \mathbb{R}$, $\forall\ s,t \in \mathbb{R}_+$, *and* $\forall\ f \in E^+$,

$$P_{s+t}f(x) = P_s(P_tf)(x), \quad Q_{s+t}f(x) = Q_s(Q_tf)(x) \text{ for } \pi_r \text{ a.e. } x.$$

(b) *Let* $r \in \mathbb{R}$ *and* $(t_n)$ *a decreasing sequence of real numbers with* $t_n \to 0$ *as* $n \to \infty$. *Then* $P_{t_1}(x,E) \leq P_{t_2}(x,E) \leq \cdots \uparrow 1$ *as* $n \to \infty$ *and*

$$Q_{t_1}(x,E) \leq Q_{t_2}(x,E) \leq \cdots \uparrow 1 \text{ as } n \to \infty \text{ for } \pi_r \text{ a.e. } x.$$

Built into the definition of a two-sided time homogeneous Markov process P is the statement: If, for each t, $\pi_t$ is the P-distribution of $X_t$ on $\{X_t \in E\}$ and $(P_t)$ and $(Q_t)$ are the forward and backward transitions, then

$$(3.5) \qquad \pi_s(f(P_{t-s}g)) = \pi_t((Q_{t-s}f)g) \quad \text{if } s \leq t \text{ and } f,g, \in E^+.$$

We might describe this condition by saying that $(P_t)$ and $(Q_t)$ are in "duality" relative to the family of measures $(\pi_t)$.

(3.6) THEOREM. *Let* $(\pi_t\colon t \in \mathbb{R})$ *be a family of* $\sigma$*-finite measures on* $(E,E)$, *and* $(P_t\colon t \in \mathbb{R}_+)$, $(Q_t\colon t \in \mathbb{R}_+)$ *families of sub-Markov kernels on* $(E,E)$ *so that* (3.4a,b) *and* (3.5) *hold. Also assume that* $(E,E)$ *is standard (i.e. isomorphic to a Borel subset of a complete separable metric space). Then there exists a two-sided time-homogeneous Markov process with forward and backward transitions* $(P_t)$ *and* $(Q_t)$ *so that* $\forall\ t \in \mathbb{R}$, *and* $\forall\ f \in E^+$, $Pf(X_t) = \pi_tf$.

PROOF. For $s < t$ define $\pi_{s,t}$ on $(E \times E, \mathcal{E} \otimes \mathcal{E})$ by: $\iint \pi_{s,t}(dx,dy) f(x,y) = \int \pi_s(dx) \int P_{t-s}(x,dy)f(x,y)\ \forall\ f \in (\mathcal{E} \otimes \mathcal{E})^+$. Now fix $f \in \mathcal{E}^+$, and $t < u$. Then $\iint \pi_{t,u}(dx,dy)f(x) = \int \pi_t(dx)f(x)P_{u-t}(x,E)$. Hence, by our hypotheses, $\iint \pi_{t,u}(dx,dy)f(x) \uparrow \pi_t f$ as $u \downarrow t$. Also for a fixed $f \in \mathcal{E}^+$ and $s < t$ we have $\iint \pi_{s,t}(dx,dy)f(y) = \int \pi_s(dx)P_{t-s}f(x) = \pi_s(1 (P_{t-s}f)) = \pi_t((Q_{t-s}1)f) = \iint \pi_t(dy)f(y)Q_{t-s}(y,E)$. Again our hypotheses imply that $\iint \pi_{s,t}(dx,dy)f(y) \uparrow \pi_t f$ as $s \uparrow t$.

We now refer the reader to [3] to see that, together with our hypotheses, the preceding paragraph shows that all the hypotheses of the main theorem of [3] are satisfied. The conclusion is the existence of a measure P which is Markov (see(3.2a)) and satisfies:

(i) $\forall\ t \in \mathbb{R}$, and $\forall\ f \in \mathcal{E}^+$, $Pf(X_t) = \pi_t f$, and

(ii) $\forall\ s < t$, and $\forall\ f,g \in \mathcal{E}^+$, $P(f(X_s)g(X_t)) = \iint \pi_{s,t}(dx,dy)f(x)g(y)$.

Thus, if $s < t$ and $f,g \in \mathcal{E}^+$, then by (3.5) we have that $P(f(X_s)g(X_t)) = \pi_s(f(P_{t-s}g)) = \pi_t((Q_{t-s}f)g)$, and the proof is complete. □

For the remainder of this section we shall only consider families $(P_t)$ and $(Q_t)$ which are representable; see (2.7). Of course, in section 2 the c in the definition of $\Omega_+^c$ (see(2.1)) is just a "dummy" symbol. Thus we shall let $(P^x : x \in E)$ be a family of probabilities on $(\Omega_+^b, \mathcal{F}_+^b)$ which represents $(P_t)$ and we shall let $(Q^x : x \in E)$ be a family of probabilities on $(\Omega_+^a, \mathcal{F}_+^a)$ which represents $(Q_t)$; see (2.7). Note that for such families the conditions (3.4a,b) hold for any choice of measures $(\pi_t)$. Thus, in addition to the hypothesis that $(E,\mathcal{E})$ be standard, the only hypothesis with content in (3.6) for such families is condition (3.5).

(3.7) REMARK. Together with some regularity assumptions, strong Markov assumptions, etc., the hypothesis for the construction of an auxiliary process in [4] is (3.5) for representable families in the case where $\pi_s = \pi_t\ \forall\ s,t$. In this case $(Q_t)$ would be written more familiarly as $(\hat{P}_t)$.

Our immediate aim is to give two-sided versions of (2.11) and (2.13). First we need some definitions and some simple properties.

(3.8) DEFINITION. Let $(P^x)$ (resp. $(Q^x)$) be a family of probabilities on $(\Omega_+^b, F_+^b)$ (resp. $(\Omega_+^a, F_+^a)$) which represents $(P_t)$ (resp. $(Q_t)$) as in (2.7). Define $\phi:\Omega_+^b \times \Omega_+^a \to \Omega$ by $\phi(\omega_1,\omega_2) = \omega$ where

$$\omega(t) = \begin{cases} \omega_1(t) \text{ if } t \geq 0 \\ \omega_2(-t) \text{ if } t < 0. \end{cases}$$

For every $x \in E$, let $(P,Q)^x = (P^x \otimes Q^x)\circ\phi^{-1}$. The family $((P,Q)^x : x \in E)$ is called the *family generated by* $(P,Q)$.

(3.9) PROPOSITION. *Let* $((P,Q)^x : x \in E)$ *be as in* (3.8). *Then*

(a) $\forall\, x \in E$, $(P,Q)^x$ *is a probability with* $(P,Q)^x(X_0 = x) = 1$.

(b) $\forall\, Y \in F^+$, $x \to (P,Q)^x Y$ is *E-measurable*.

PROOF. (a) follows easily by definition. Also a monotone class argument shows that (b) need only be checked for $Y = f_1(X_{t_1})\cdots f_n(X_{t_n})$ where $-\infty < t_1 < \cdots < t_n < \infty$ and $f_1,\ldots,f_n \in E^+$. For such a Y the value of $(P,Q)^x Y$ of course depends on where 0 is in relation to the set $\{t_1,\ldots,t_n\}$. For instance, suppose $t_1 \leq 0 < t_2$. In this case, by definition, $(P,Q)^x Y = Q^x f_1(X_{-t_1}) P^x(f_2(X_{t_2})\cdots f_n(X_{t_n}))$, which is $E$-measurable as a function of x by (2.8d). The argument is similar for the other possible locations of 0. The proof is complete.

(3.10) DEFINITION. Let $(M^x : x \in E)$ be a family of probabilities on $(\Omega, F)$ so that $M^x(X_0 = x) = 1 \;\; \forall\, x \in E$, and $x \to M^x Y$ is $E$-measurable $\forall\, Y \in F^+$. Also, let M be a measure on $(\Omega, F)$ so that for every $t \in \mathbb{R}$ there exists a $\sigma$-finite measure $\mu_t$ on $(E, E)$ so that, $\forall\, Y \in E^+$, $Mf(X_t)$

$= \mu_t f$. We say that M is *subordinate to the family* $(M^x)$ if $\forall\, Y \in F^+$ and $\forall\, t \in \mathbb{R}$ we have that

$$M(Y\circ\theta_t \mid \sigma(X_t)) = M^{X_t} Y \text{ on } \{X_t \in E\}.$$

(3.11) THEOREM. *Let* P *be a two-sided time-homogeneous Markov process with forward and backward transitions* $(P_t)$ *and* $(Q_t)$. *Further, suppose* $(P_t)$ *and* $(Q_t)$ *are representable. Then* P *is subordinate to* $((P,Q)^x)$.

PROOF. Fix t. By a monotone class argument the result will follow if we can show that

$$P(g(X_t)\ Y\circ\theta_t) = P(g(X_t)[(P,Q)^{X_t} Y]$$

whenever $g \in E^+$ with $Pg(X_t) < \infty$ and Y has the form $Y = f_1(X_{t_1})\cdots f_n(X_{t_n})$ where $-\infty < t_1 < t_2 < \cdots < t_n < \infty$ and $f_1,\ldots,f_n \in E^+$.

As in the proof of (3.9) the position of 0 plays a key role. For instance, if $t_1 \le 0 < t_2$, then $\forall\, x$ we have

$$(P,Q)^x Y = Q^x f_1(X_{-t_1}) P^x (f_2(X_{t_2})\cdots f_n(X_{t_n}))$$

$$= [Q_{-t_1} f_1(x)][P_{t_2}(f_2 P_{t_3-t_2}(\cdots(f_{n-1}P_{t_n-t_{n-1}} f_n)\cdots))(x)].$$

Hence

$$P(g(X_t)[(P,Q)^{X_t} Y])$$

$$= \pi_t(g(Q_{-t_1} f_1)(P_{t_2}(f_2 P_{t_3-t_2}(\cdots(f_{n-1}P_{t_n-t_{n-1}} f_n)\cdots)))).$$

On the other hand, $Y\circ\theta_t = f_1(X_{t_1+t})\cdots f_n(X_{t_n+t})$, and by (3.3),

$$P(g(X_t)\ Y\circ\theta_t) = P(g(X_t)Q_{-t_1}f_1(X_t)f_2(X_{t_2+t})\cdots f_n(X_{t_n+t}))$$

$$\vdots$$

$$= \pi_t(g(Q_{-t_1}f_1)P_{t_2}(f_2P_{t_3-t_2}(\cdots(f_{n-1}P_{t_n-t_{n-1}}f_n)\cdots))),$$

and thus for this case $P(g(X_t)\ Y\circ\theta_t) = P(g(X_t)[(P,Q)^{X_t}Y])$. The proof for the other possible positions of 0 is similar, and we are done.

In contrast to the one-sided theory (see (2.12)) it is rarely the case that each $(P,Q)^x$ is itself time-homogeneous with transitions $(P_t)$ and $(Q_t)$. The following result shows exactly when this can happen.

(3.12) THEOREM. *Let* $(M^x : x \in E)$ *be a family of probabilities on* $(\Omega, F)$ *so that* $M^x(X_0 = x) = 1\ \forall\ x \in E$, *and* $x \to M^x Y$ *is* $E$*-measurable* $\forall\ Y \in F^+$. *Suppose that* $(M^x)$ *is self-subordinate in the sense that* $\forall\ y \in E$, $M^y$ *is subordinate to* $(M^x)$ *according to* (3.10). *Then there exists a family* $(\tau_t : t \in \mathbb{R}_+)$ *of functions from* $E \cup \{b\}$ *to* $E \cup \{b\}$ *and a family* $(\gamma_t : t \in \mathbb{R}_+)$ *of functions from* $E \cup \{a\}$ *to* $E \cup \{a\}$ *satisfying:*

(i) $\forall\ t \in \mathbb{R}_+$, $\tau_t(b) = b$ *and* $\gamma_t(a) = a$.

(ii) $\forall\ s,t \in \mathbb{R}_+$, $\tau_{s+t} = \tau_t \circ \tau_s$ *and* $\gamma_{s+t} = \gamma_s \circ \gamma_t$. *Also* $\tau_0$ *is the identity on* $E \cup \{b\}$, *and* $\gamma_0$ *is the identity on* $E \cup \{a\}$.

(iii) $\forall\ x,y \in E$, *and* $\forall\ t \in \mathbb{R}_+$, $\tau_t(x) = y \iff \gamma_t(y) = x$.

(iv) $\forall\ x \in E$, *and* $\forall\ t \in \mathbb{R}_+$, $X_t = \tau_t(x)$ *and* $X_{-t} = \gamma_t(x)$ $M^x$-a.s.

PROOF. Let $x \in E$, $f \in E^+$, and $t \in \mathbb{R}_+$. Then

$$M^x f(X_t) = M^x(f(X_t); X_0 = x)$$

$$= M^x(f(X_t); \theta_t^{-1}(X_{-t} = x))$$

$$= M^x(f(X_t)M^{X_t}(X_{-t} = x)).$$

It follows that $\forall\, x \in E$ and $\forall\, t \in \mathbb{R}_+$, $M^{X_t}(X_{-t} = x) = 1$ $M^x$-a.s. on $\{X_t \in E\}$. Thus we have

(3.13) $\forall\, x \in E$ and $\forall\, t \in \mathbb{R}_+$, either $M^x(X_t = b) = 1$ or there exists a $y \in E$ with $M^y(X_{-t} = x) = 1$.

Similarly by considering $M^x f(X_{-t})$ instead of $M^x f(X_t)$ in the above argument, we have a dual fact:

(3.14) $\forall\, x \in E$ and $\forall\, t \in \mathbb{R}_+$, either $M^x(X_{-t} = a) = 1$ or there exists a $y \in E$ with $M^y(X_t = x) = 1$.

Now suppose $x,y \in E$ with $M^y(X_{-t} = x) = 1$. Then

$$1 = M^y(X_{-t} = x;\ X_0 = y) = M^y(X_{-t} = x;\ \theta^{-1}_{-t}(X_t = y))$$

$$= M^y(M^{X_{-t}}(X_t = y);X_{-t} = x) = M^y(X_{-t} = x)M^x(X_t = y).$$

Thus $M^x(X_t = y) = 1$. Similarly, if $x,y \in E$ with $M^y(X_t = x) = 1$, then $M^x(X_{-t} = y) = 1$

For each $t \in \mathbb{R}_+$ define $\tau_t\colon E \cup \{b\} \to E \cup \{b\}$ by $\tau_t(b) = b$, and $\forall\, x \in E$, $\tau_t(x) = y$ if $M^x(X_t = y) = 1$. Also, define $\gamma_t\colon E \cup \{a\} \to E \cup \{a\}$ by $\gamma_t(a) = a$, and $\forall\, x \in E$, $\gamma_t(x) = y$ if $M^x(X_{-t} = y) = 1$. (The preceding paragraph justifies these definitions.) Clearly $\tau_0$ is the identity on $E \cup \{b\}$ and $\gamma_0$ is the identity on $E \cup \{a\}$.

Now, fix $s,t \in \mathbb{R}_+$. By definition, $\tau_{s+t}(b) = \tau_s \circ \tau_t(b) = b$. Let $x \in E$, $y = \tau_t(x)$, and $z = \tau_{s+t}(x)$. We want to show that $\tau_s(y) = z$. By definition of $\Omega$, if $y = b$ then $z = b$ and thus $\tau_s(y) = z$. Next, suppose that $y \in E$. Then

$$1 = M^x(X_{s+t} = z) = M^x(X_t = y, X_{s+t} = z)$$

$$= M^x(M^{X_t}(X_s = z); X_t = y) = M^x(X_t = y)M^y(X_s = z)$$

and thus $M^y(X_s = z) = 1$, which means that $\tau_s(y) = z$. Thus $\tau_{s+t} = \tau_s \circ \tau_t$. A similar argument shows that $\gamma_{s+t} = \gamma_s \circ \gamma_t$.

Let $x, y \in E$, $t \in \mathbb{R}_+$, and suppose $\tau_t(x) = y$. Then

$$1 = M^x(X_t = y) = M^x(X_0 = x, X_t = y)$$

$$= M^x(M^{X_t}(X_{-t} = x); X_t = y) = M^x(X_t = y)M^y(X_{-t} = x)$$

which gives that $M^y(X_{-t} = x) = 1$, or $\gamma_t(y) = x$. Similarly $\gamma_t(y) = x \Rightarrow \tau_t(x) = y$.

Finally, if $x \in E$, $t \in \mathbb{R}_+$, then by definition $X_t = \tau_t(x)$ and $X_{-t} = \gamma_t(x)$ $M^x$-a.s. □

(3.15) REMARKS. (a) We may say that in (3.12) each $M^x$ determines a deterministic flow with absorption to the right at b and with absorption to the left at a.

(b) The significance of the condition (3.12iii) in relation to the other conditions is that $(M^x)$ is the family generated (see (3.8)) by $(P,Q)$ where $\forall\, t \in \mathbb{R}_+$ and $\forall\, f \in \mathcal{E}^+$, $P_t f(x) = M^x f(X_t)$ and $Q_t f(x) = M^x f(X_{-t})$, and that each $M^x$ is time-homogeneous Markov with transitions $(P_t)$ and $(Q_t)$; we omit the details. This leads to the next result.

(3.16) COROLLARY. *Let* $(P_t)$ *and* $(Q_t)$ *be representable families and suppose that* $\forall\ x \in E$, $(P,Q)^x$ *is time-homogeneous Markov with forward and backward transitions* $(P_t)$ *and* $(Q_t)$. *Then there exists a family* $(\tau_t : t \in \mathbb{R}_+)$ *of functions from* $E \cup \{b\}$ *to* $E \cup \{b\}$ *and a family of functions* $(\gamma_t : t \in \mathbb{R}_+)$ *from* $E \cup \{a\}$ *to* $E \cup \{a\}$ *satisfying* (i), (ii), *and* (iii) *of* (3.12) *and such that* $\forall\ x \in E$ *and* $\forall\ f \in \mathcal{E}^+$, $P_t f(x) = f(\tau_t(x))$ *and* $Q_t f(x) = f(\gamma_t(x))$.

PROOF. According to (3.11), if we let $M^x = (P,Q)^x$ for every $x \in E$ then $(M^x)$ satisfies the hypotheses of (3.12) and the desired result follows from (3.12) and the definition of the family $((P,Q)^x)$.

Thus, the two-sided theory is analogous to the one-sided theory through the concept of subordination, but the analogy ends when one considers self-subordinate families. Another explanation may be as follows:

If $(P_t)$ is a representable family, then one only needs the family $(P^x)$, which represents $(P_t)$, and an arbitrary initial measure to describe all one-sided time-homogeneous Markov processes with transitions $(P_t)$. However, if $(Q_t)$ is another representable family, then to describe all two-sided time-homogeneous Markov processes with forward and backward transitions $(P_t)$ and $(Q_t)$ one needs the family $((P,Q)^x)$ and a *family* of measures $(\pi_t)$, serving as the one-dimensional distributions, which must be related to $(P_t)$ and $(Q_T)$ according to the constraint (3.5). Thus, in the two-sided case a certain degree of freedom is lost in that, in general, there is nothing analogous to the selection of an arbitrary initial measure for one-sided processes.

The next section addresses this problem by considering a certain special class of two-sided processes.

## 4. Quasi-Stationary Two-Sided Processes

As seen in section 3, given families $(P_t)$ and $(Q_t)$ which are representable (and thus are semigroups) it does not, in general, seem possible to construct a two-sided time-homogeneous Markov process with forward and backward transitions $(\dot{P}_t)$ and $(Q_t)$ and with a pre-assigned measure as its distribution at a fixed time, say $t = 0$. The purpose of this section is to give a sufficient condition on a pre-assigned measure that will ensure that it can be the distribution of $X_0$ for some two-sided process. For this, we only consider the forward transitions $(P_t)$ as given.

(4.1) DEFINITION. Let $(P_t)$ be a family of sub-Markov kernels on $(E,\mathcal{E})$, where $(E,\mathcal{E})$ is as in section 2. Also let $\lambda \in \mathbb{R}$, and $\pi$ be a $\sigma$-finite measure on $(E,\mathcal{E})$. We call $\pi$ *$\lambda$-excessive relative to* $(P_t)$ if $e^{-\lambda t}\pi P_t \leq \pi$ $\forall\, t \in \mathbb{R}_+$, and $e^{-\lambda t}\pi P_t \uparrow \pi$ as $t \downarrow 0$.

(4.2) REMARKS. (a) This definition, of course, accords with the usual definition of excessiveness if $\lambda \geq 0$ and $(P_t)$ is a semigroup.
(b) If $\pi$ is a $\sigma$-finite measure and $(P_t)$ is a semigroup with $e^{-\lambda t}\pi P_t \leq \pi$ $\forall\, t \in \mathbb{R}_+$, then it is easy to check that $\pi' = \lim_{t\to 0} e^{-\lambda t}\pi P_t$ is $\lambda$-excessive relative to $(\dot{P}_t)$.

(4.3) THEOREM. *Assume* $(E,\mathcal{E})$ *is standard; see* (3.6). *Let* $(P_t)$ *be a family of sub-Markov kernels on* $(E,\mathcal{E})$, $\pi$ *a $\sigma$-finite measure on* $(E,\mathcal{E})$, *and* $\lambda \in \mathbb{R}$ *so that* $\pi$ *is $\lambda$-excessive relative to* $(P_t)$. *Further suppose:*

(i) $\forall\, s,t \in \mathbb{R}_+$ *and* $\forall\, f \in \mathcal{E}^+$, $P_{s+t}f(x) = P_s(P_tf)(x)$ *for* $\pi$ *a.e.* $x$, *and*

(ii) $\forall$ *sequence* $(t_n)$ *so that* $t_n \downarrow 0$ *as* $n \uparrow \infty$, $P_{t_1}(x,E) \leq P_{t_2}(x,E) \leq \cdots \uparrow 1$ *as* $n \uparrow \infty$ *for* $\pi$ *a.e.* $x$.

*Then there exists a two-sided time-homogeneous Markov process* P *with*

$(P_t)$ *as forward transitions and so that* $Pf(X_0) = \pi f \ \forall f \in E^+$, and $P \circ \theta_t^{-1} = e^{\lambda t} P \ \forall t \in \mathbb{R}$.

PROOF. For $t \in \mathbb{R}$ define $\pi_t = e^{\lambda t}\pi$, and for $s < t$ define $\pi_{s,t}$ on $(E \times E, E \otimes E)$ by

$$\iint \pi_{s,t}(dx,dy) f(x,y) = \int \pi_s(dx) \int P_{t-s}(x,dy) f(x,y) \ \forall f \in (E \otimes E)^+.$$

If $f \in E^+$ and $t < u$, then

$$\iint \pi_{t,u}(dx,dy) f(x) = \int \pi_t(dx) f(x) P_{u-t}(x,E).$$

By (ii) it follows that $\iint \pi_{t,u}(dx,dy) \uparrow \pi_t f$ as $u \downarrow t$.

Next, if $f \in E^+$, and $s < t$, then

$$\iint \pi_{s,t}(dx,dy) f(y) = \int \pi_s(dx) P_{t-s} f(x) = e^{\lambda t}(e^{-\lambda(t-s)} \pi P_{t-s}(f)).$$

Thus, since $\pi$ is $\lambda$-excessive relative to $(P_t)$ then

$$\iint \pi_{s,t}(dx,dy) f(y) \uparrow e^{\lambda t} \pi f = \pi_t f \text{ as } s \uparrow t.$$

Since $\forall r \in \mathbb{R}$, $\forall s,t \in \mathbb{R}_+$, and $\forall f \in E^+$ we have, by the definition of $\pi_r$ and (i), $P_{s+t} f(x) = P_s(P_t f)(x)$ for $\pi_r$ a.e. $x$, then the hypotheses of the main theorem of [3] are satisfied and thus there exists a two-sided Markov process P with

(i) $\forall t \in \mathbb{R}$ and $\forall f \in E^+$, $Pf(X_t) = e^{\lambda t} \pi f$, and

(ii) $\forall s < t$, and $\forall f,g \in E^+$, $P(f(X_s) g(X_t)) = \iint \pi_{s,t}(dx,dy) f(x) g(y)$.

Fix $t \in \mathbb{R}$, $t_1 < t_2 < \cdots < t_n$, and $f_1, \ldots, f_n \in E^+$. Then

$$P(f_1(X_{t_1})\cdots f_n(X_{t_n}))$$

$$= e^{\lambda t_1}\pi(f_1P_{t_2-t_1}(f_2P_{t_3-t_2}(\cdots(f_{n-1}P_{t_n-t_{n-1}}f_n)\cdots))).$$

Also,

$$P([f_1(X_{t_1})\cdots f_n(X_{t_n})]\circ\theta_t)$$

$$= P(f_1(X_{t+t_1})\cdots f_n(X_{t+t_n}))$$

$$= e^{\lambda(t+t_1)}\pi(f_1P_{t_2-t_1}(f_2P_{t_3-t_2}(\cdots(f_{n-1}P_{t_n-t_{n-1}}f_n)\cdots)))$$

$$= e^{\lambda t}P(f_1(X_{t_1})\cdots f_n(X_{t_n}))\,;$$

A standard monotone class argument gives that $P\circ\theta_t^{-1} = e^{\lambda t}P \ \forall\ t \in \mathbb{R}$.

Finally, let $\forall\ t \in \mathbb{R}_+$, $Q_t$ be a sub-Markov kernel on $(E,\mathcal{E})$ so that $\forall\ f,g \in \mathcal{E}^+$, $P(f(X_{-t})g(X_0)) = P(Q_tf(X_0)g(X_0))$. Such a kernel exists by the usual construction of a regular conditional distribution, using the fact that $\pi$ is $\sigma$-finite and $(E,\mathcal{E})$ is a standard space. It now follows that $\forall\ s < t$ and $\forall\ f,g \in \mathcal{E}^+$,

$$\pi_s(f(P_{t-s}g)) = P(f(X_s)g(X_t)) = P([f(X_{-(t-s)})g(X_0)]\circ\theta_t)$$

$$= e^{\lambda t}P(f(X_{-(t-s)})g(X_0)) = e^{\lambda t}P(Q_{t-s}f(X_0)g(X_0))$$

$$= e^{\lambda t}\pi((Q_{t-s}f)g) = \pi_t((Q_{t-s}f)g).$$

According to (3.2), P is a two-sided time-homogeneous Markov process with forward and backward transitions $(P_t)$ and $(Q_t)$. □

(4.4) REMARK. In the preceding theorem, if $\lambda = 0$ then $\pi$ is excessive relative to $(P_t)$ and the resulting process P is stationary in time. In this case we say that $(P_t)$ and $(Q_t)$ are in weak duality relative to $\pi$. For arbitrary $\lambda$ we call the P of (4.3) a *quasi-stationary Markov process*.

## References


1. B.W. ATKINSON. "Two-sided Markov chains." Preprint (1984).

2. K.L. CHUNG and J.B. WALSH. "To reverse a Markov process." *Acta. Math.* 123, 225-251 (1969).

3. S.E. KUZNETSOV. "Construction of Markov processes with random times of birth and death." *Th. Prob. Appl. 18*, 571-575 (1973).

4. J.B. MITRO. "Dual Markov processes: Construction of a useful auxiliary process." *Z. Wahr. verw. Geb. 47*, 139-156 (1979).

5. D.W. STROOCK. "On the spectrum of Markov semi-groups and the existence of invariant measures." *Functional Analysis in Markov Processes*, Lecture notes in Math. 923, pp. 286-307. Springer-Verlag, New York, 1982.


B.W. ATKINSON
Department of Mathematics
University of Florida
Gainesville, Florida 32611

*Seminar on Stochastic Processes, 1984*
Birkhäuser, Boston, 1986

# THE BEHAVIOUR AND CONSTRUCTION OF LOCAL TIMES FOR LÉVY PROCESSES

by

MARTIN T. BARLOW[1], EDWIN A. PERKINS, S. JAMES TAYLOR[1]

## 1. Introduction

The local time of a Lévy process $X_t$ at a point $a$, denoted $\ell_t^a(X)$, is a continuous increasing additive functional, which increases only on $\{t: X_t = a\}$. If $X$ is such that $\ell_t^0$ exists, then as the transition probabilities of $X$ are stationary in space, $\ell_t^x$ will exist for every $x \in \mathbb{R}$, and we may therefore ask about the properties of the map $(x,t,\omega) \to \ell_t^x(\omega)$.

In this paper we give a survey of what is known about this problem, and include some new results of the authors. After establishing our notation in section 2, we review in section 3 known conditions for the existence of a jointly continuous local time, and the properties of $\ell_t^x$ when a continuous version does not exist. We present a conjecture of J. Hawkes, which gives necessary and sufficient conditions for the existence of a continuous version of $(x,t) \to \ell_t^x$, and formulate some other problems concerning its behaviour.

In section 4 we look at the case when the range of $X$ is nowhere dense: this forces $\ell_t^x$ to have a very erratic behaviour, and in par-

[1]Most of the research leading to this paper was carried out while these authors were visiting the University of British Columbia with support from N.S.E.R.C.

ticular we show that, if $\ell_t^x$ is unbounded, then $\ell_t^x$, $x \in \mathbb{R}$, is dense in $[0,\infty)$.

In the final two sections we consider the problem of constructing $\ell_t^x$ as the limit of a sequence $K_n(x,t)$ of functionals of the path of $X_t$. If $\lim_{n\to\infty} K_n(0,t) = \ell_t^0$ a.s., then, using Fubini, we have immediately that $K_n(x,t)$ converges to $\ell_t^x$ on a set of full Lebesgue measure; but to go further than this requires new techniques.

In section 5 we give some examples of constructions which, while converging almost everywhere, fail to converge to $\ell_t^x$ at some levels: some of these counterexamples are valid for Brownian motion.

Finally, in section 6, we state three positive results on the uniform convergence in x of specific constructions $K_n(x,t)$ to $\ell_t^x$. One of these is proved here; the proofs of the others are rather complicated, and will be given in a subsequent paper [4].

## 2. Preliminaries

We use the framework established by Getoor and Kesten [13] which combines the definition of local time at a fixed level as a continuous additive functional with its definition as an occupation density in state space. In this paper a Lévy process $X_t$ will be a standard Markov process on the line with stationary independent increments whose characteristic function takes the form

$$E\, e^{i\lambda X_t} = E\, e^{i\lambda(X_{t+s}-X_s)} = e^{-t\psi(\lambda)}$$

with

(1) $$\psi(\lambda) = -ia\lambda + \tfrac{1}{2}\sigma^2\lambda^2 - \int_{R\setminus\{0\}} \left[e^{i\lambda y} - 1 - \frac{i\lambda y}{1+y^2}\right]\nu(dy).$$

As shown by Kesten [17] and Bretagnolle [7], if either $\sigma^2 > 0$, or

(2) $$\int (|x| \wedge 1)\nu(dx) = \infty \quad \text{and} \quad \int_0^\infty \mathbb{R}\mathrm{e}\,\frac{1}{1+\psi(\lambda)}\,d\lambda < \infty,$$

then 0 is regular for both $\{0\}$ and $(\mathbb{R}\backslash\{0\})$, and the local time $\ell_t^x$ exists for fixed x as a continuous additive functional in t. It will be convenient to have a notation for particular processes satisfying these conditions.

$B_t$ denotes a standard Brownian motion for which $\psi(\lambda) = \frac{1}{2}\lambda^2$.

$S_t^{\gamma,\delta}$ denotes a stable process of index $\gamma$ with

$$\psi(\lambda) = |\lambda|^\gamma (1 + i\delta \tan \tfrac{\pi\alpha}{2}) \qquad -1 \le \delta \le 1, \quad 1 < \gamma < 2$$

arising from a Lévy measure of the form

$$\nu(dx) = c|x|^{-1-\gamma}(pI_{(x>0)} + qI_{(x<0)})$$

with $p \ge 0,\ q \ge 0,\ p+q = 1,\ p-q = \delta.$

$S_t^{1,\alpha,\beta}$ is a symmetric process close to Cauchy with

$$\nu(dx) = x^{-2}\, g^{\alpha,\beta}\left(\frac{1}{|x|}\right) \quad \text{and}$$

$$g^{\alpha,\beta}(y) = (\log y)^\alpha (\log\log y)^\beta I_{(y>e)}: \quad \alpha,\beta \in \mathbb{R}.$$

$A_t^{1,\alpha,\beta,p}$ is the corresponding asymmetric process with

$$\nu(dx) = x^{-2}\, g^{\alpha,\beta}\left(\frac{1}{|x|}\right)(pI_{(x>0)} + qI_{(x<0)})$$

and $p \ge 0,\ q \ge 0,\ \ p+q = 1,\ \ p \ne \frac{1}{2}.$

It is clear that (2) is satisfied for $S_t^{\gamma,\delta}$, $1 < \gamma < 2$ and Barlow [2] estimates $\psi(\lambda)$ and shows that (2) is satisfied for

$$S^{1,\alpha,\beta} \quad \text{if and only if} \quad \alpha > 1 \quad \text{or} \quad \alpha = 1,\ \beta > 1$$

$$A^{1,\alpha,\beta,p} \text{ if and only if } \alpha > -1 \text{ or } \alpha = -1, \ \beta > 1.$$

We now summarise the main content of Theorem 4 in [13].

THEOREM 2.1. *Suppose* $X_t$ *is a Lévy process whose exponent* (1) *satisfies either* (2) *or* $\sigma^2 > 0$. *Then, for any* $r > 0$ *there exists a bounded continuous density* $u^r$ *for the potential kernel; that is,*

(3) $$E^x \int_0^\infty e^{-rt} f(X_t)dt = \int_{\mathbb{R}} u^r(y-x)f(y)dy$$

*for each non-negative measurable* f. *For each* x *there exists a continuous additive functional* $\ell_t^x$ *(a local time at* x*) such that*

(4) $$E^x \int_0^\infty e^{-rt} d_t \ell_t^y = u^r(y-x),$$

*and for fixed* $t \geq 0$, *the map* $(x,\omega) \to \ell_t^x$ *is* $\mathcal{B} \times \mathcal{F}_t$ *measurable, and a.s. for each Borel set* B,

(5) $$\mu_t(B) = |\{s \leq t : X_s \in B\}| = \int_B \ell_t^x \, dx.$$

*The probability that* $\ell_t^x$ *has a version continuous in* $(x,t)$ *is zero or one.*

## 3. Continuity of local time

In general the conditions satisfied by $\ell_t^x$ in Theorem 2.1 do not determine it uniquely as a function of $(x,t)$. However, if the process is such that a.s. a continuous version exists, it is clear by (5) that this version is unique. We will denote it by $L_t^x$. It is instructive to see how we can modify $L_t^x$ to obtain a new version which is not continuous, but still satisfies Theorem 2.1.

Suppose $Q_t(\omega)$ is a random subset of $\mathbb{R}$ such that, for each $t > 0$,

$$I_{Q_t}(x) \text{ is } \mathcal{B}\times\mathcal{F}_t \text{ measurable,}$$

$$t' > t \Rightarrow Q_{t'} \supset Q_t,$$

$$Q_t(\omega) \text{ is dense in } \{X_s \mid s \leq t\},$$

(6) $$P\{x \in Q_t(\omega)\} = 0 \text{ for all } x.$$

For the process $B_t$ we give an example of such a set $Q_t$ in section 5 (see Example 5.4). Now define

$$\ell_t^x = L_t^x \text{ for all } t \geq 0 \text{ whenever } x \notin Q = \bigcup_{t\geq 0} Q_t,$$

for $x \in Q$, let $t_0(x) = \inf\{t : x \in Q_t\}$ and let

$$\ell_t^x = L_t^x \text{ for } t \leq t_0; \quad \ell_t^x = L_{t_0}^x \text{ for } t > t_0.$$

It is easy to check that $\ell_t^x$ satisfies all the conditions of Theorem 2.1 and for each $x$, $\ell_t^x$ is monotone and continuous in $t$. However, if $x_0 \in Q_t(\omega)$ for $t > t_1$ and $t_1$ is a growth point of $L_t^{x_0}$, then $\ell_t^x$ will be discontinuous in $x$ at $x = x_0$ for $t > t_1$. By (6) we have still, for each $(x,t)$, $\ell_t^x = L_t^x$ a.s.

In fact, without any continuity assumption, the normalisation (4) ensures that any two versions of $\ell_t^x$ will agree a.s. for all $t > 0$ and a fixed level $x$. This agreement therefore extends a.s. to all levels $x$ in a fixed countable set D. We will assume that D is the set of dyadic rationals. We can then study the a.s. properties of any version $\ell_t^x$ satisfying Theorem 2.1 by looking at its behaviour for $x \in D$, $t > 0$. For example, to show that a jointly continuous version of $\ell_t^x$ exists, it suffices to show that $\{\ell_t^x,\ 0 \leq t \leq 1,\ x \in D\}$ is uniformly continuous.

Necessary and sufficient conditions for the existence of a continuous version of $\ell_t^x$ are not known. Sufficient conditions have been given by Trotter [31], Boylan [6], Getoor and Kesten [13], and Barlow [2]. Getoor and Kesten also found a condition which ensures that no continuous version of $\ell_t^x$ exists: this last result was strengthened by Millar and Tran [22], who showed that, under the same conditions, $\ell_t^x$ is a.s. unbounded.

For the special processes introduced earlier, we have the following table.

| Process | Parameter Values | Properties of Local Time |
|---|---|---|
| $S^{\alpha,\delta}$ | $1 < \gamma < 2$, $-1 \le \delta \le 1$ | continuous |
| $S^{1,\alpha,\beta}$ | $\alpha > 2$; $\alpha = 2$, $\beta > 2$ | continuous |
| | $\alpha = 2$, $0 < \beta \le 2$ | unknown |
| | $\alpha = 2$, $\beta \le 0$; $1 < \alpha < 2$; $\alpha = 1$, $\beta > 1$ | unbounded on D |
| $A^{1,\alpha,\beta,p}$ | $\alpha > 0$; $\alpha = 0$, $\beta > 2$ | continuous |
| | $\alpha = 0$, $0 < \beta \le 2$ | unknown |
| | $\alpha = 0$, $\beta \le 0$; $-1 < \alpha < 0$; $\alpha = -1$, $\beta > 1$ | unbounded on D |

If the following conjecture is correct, then the local times of $S^{1,2,\beta}$, $A^{1,0,\beta,p}$, $0 < \beta \le 2$ are not continuous.

CONJECTURE 3.1 (Hawkes, 1981). *Let*

$$\phi^2(h) = \frac{1}{\pi} \int (1 - \cos \lambda h) \operatorname{Re}\Big(\frac{1}{1+\psi(\lambda)}\Big)\, d\lambda,$$

*and let* $\bar{\phi}$ *be the monotone rearrangement of* $\phi$. *Set*

$$I(\bar{\phi}) = \int_{0+} \frac{\bar{\phi}(u)}{u(\log 1/u)^{\frac{1}{2}}}\, du.$$

*Then* $I(\bar{\phi}) < \infty$ *is a necessary and sufficient condition for the existence of a continuous version of* $(x,t) \to \ell_t^x$.

The sufficient condition for continuity given in Barlow [2] is that

$I(\tilde{\phi}) < \infty$, where $\tilde{\phi}(h) = \sup_{|u|\le h} \phi(u)$.

A related question is the following

PROBLEM 3.2. *If no jointly continuous version of* $\ell_t^x$ *exists, is every version of* $\ell_t^x$ *unbounded for* $t = t_0$, $x \in \mathbb{R}$?

Based on all known examples it seems possible that the process $\ell_{t_0}^x$, for fixed $t_0 > 0$, exhibits the same sort of dichotomy in behaviour as a stationary Gaussian process.

CONJECTURE 3.3. *A Lévy process satisfying* (2) *either a.s. has a continuous local time, or a.s. every version* $\ell_t^x$ *of the local time has the property that, for* $t_0 > 0$, *the values of* $\ell_{t_0}^x$, $x \in \mathbb{R}$, *are dense in* $[0,\infty)$.

We present further evidence in support of this conjecture in the next section.

We remark on another consequence of the improved modulus of continuity in space obtained by Barlow [2]. Hawkes [15] obtained an exact uniform modulus of continuity in t for fixed x for $L_t^x$, the continuous local time of the stable process $S_t^{\gamma,\delta}$. Perkins [25] has obtained the best modulus in t which is true uniformly in x:

$$\lim_{s\downarrow 0} \sup_{0\le t\le 1} \sup_{a\in\mathbb{R}} (L_{t+s}^a - L_t^a)\phi_\gamma(s)^{-1} = \theta_0,$$

where $\phi_\gamma(s) = s^{1-1/\gamma}(\log 1/s)^{1/\gamma}$, and $\theta_0$ is a known constant, strictly larger than that of Hawkes [15].

Except for $B_t$, where one can use the Ray-Knight theorem (see Ray [27] or Knight [30]), the exact modulus in space is not known. This gives

PROBLEM 3.4. *What is the asymptotic behaviour of*

$$W(y) = \sup_{x\in\mathbb{R}} \sup_{|h|\le y} |L_t^{x+h}(S^{\gamma,\delta}) - L_t^x(S^{\gamma,\delta})|,$$

*as* $y \downarrow 0$ ?

Barlow [2] obtains

$$W(y) \le c(\sup_{x\in\mathbb{R}} L_t^x)^{\frac{1}{2}} y^{\frac{1}{2}(\gamma-1)} (\log 1/y)^{\frac{1}{2}},$$

which is likely to be the right order of magnitude, since it is for $B_t$.

## 4. Processes with a nowhere dense range

We denote the range up to time $t$ by

$$F_t = \{x \in \mathbb{R} : X_s = x \quad \text{for some} \quad s \in [0,t]\}.$$

As remarked in Pruitt, Taylor [26], if $X_t$ is a Lévy process with a local time, then a.s. $F_t$ is a closed subset of $\mathbb{R}$ with positive Lebesgue measure for $t > 0$. The zero-one law of Barlow [1] shows that *either* a.s. $F_t$ is a countable union of disjoint closed intervals; *or* a.s. $F_t$ is a perfect nowhere dense set of positive Lebesgue measure. Both cases can arise. In fact Kesten [18] showed that for $S^{1,\alpha,\beta}$ we have $F_t$ nowhere dense when $\alpha = 1$, $1 < \beta < 2$ and Pruitt, Taylor [26] show that the asymmetric Cauchy process $A_t^{1,0,0,p}$ has this property except for the extreme case $p = 1$ or $0$ where jumps in only one direction occur. This leads naturally to the next

PROBLEM 4.1. *Suppose* $X_t$ *is a Lévy process satisfying* (2), *and in* (1), $\nu(-\infty,0) = +\infty = \nu(0,+\infty)$. *If a.s. no continuous version of* $\ell_t^x$ *ex-*

*ists, does it follow that the range* $F_t$ *is a.s. nowhere dense?*

Note that a nowhere dense $F_t$ implies $\ell_t^x$ is discontinuous. We cannot omit the extra condition on $\nu$ in Problem 4.1 because, if $\nu(-\infty,0)$ is finite, the sample paths of $X_t$ exhibit a local one-sided continuity which forces $F_t$ to be a union of intervals.

The following Proposition shows that, for the asymmetric Cauchy process $A^{1,0,0,p}$, $p<1$, $\{\ell_t^x, x \in \mathbb{R}\}$ is dense in $[0,\infty)$.

PROPOSITION 4.2. *Suppose* $X_t$ *is a Lévy process with local time* $\ell_t^x$ *such that*

(7) (i) The range $F_t$ is nowhere dense in $\mathbb{R}$,

(8) (ii) $\ell_t^x$ is unbounded for $x \in D \cap (-\eta,\eta)$ for all $\eta > 0$.

Then, for any interval $(a,b)$ and $t>0$, *either*

(iii) $F_t \cap (a,b) = \phi$; *or*

(iv) for all $0 \le u < v < +\infty$ there exists $y \in D \cap (a,b)$ such that $u < \ell_t^y < v$.

PROOF: The idea is to search for the point $y$ as the value of a process $Y_s$ which moves on $D$ to satisfy $u<\ell_s^{Y_s} < v$ for all $s$ greater than the hitting time of $(a,b)$. Let $u < u_0 < v_0 < v$ and put

$$\Gamma = \{(\omega,s):\ \ell_s^{X_s}(\omega) = u_0,\ X_s \in D \cap (a,b)\}.$$

Now suppose $S<T$ are any stopping times such that $X_s \in (a,b)$ for $S \le s < T$. Then

(9) $$P\{\omega:\ (\omega,s) \in \Gamma \quad \text{for some} \quad s \in (S,T)\} = 1.$$

To see this first note that (7) implies that $F_S$ is nowhere dense, so there exists a sequence $A_n$ of $\mathcal{F}_S$-measurable random variables with $A_n \notin F_S$, and $A_n \to X_S$. But then $X_S$ is regular for the set $\{A_n, n \geq 1\}$, since $P^x(y \in F_t) \to 1$ as $y \to x$, and therefore there exists $n = N$ such that $A_N \in F_T$. But now (8) implies that $\ell_T^a$, $a \in D$, is unbounded in every interval around $A_N$, and therefore, for some $Y \in D$, $\ell_T^Y > v_0$ while $Y \notin F_S$. By continuity in t we can find s with $S < s < T$, $X_s = Y$, and $\ell_S^Y = u_0$. This proves (9).

Now put $T_0 = \inf\{s \geq 0: X_s \in (a,b)\}$. Fix $\varepsilon > 0$, We shall define an optimal process Y such that for $s > T_0 + \varepsilon$, $Y_s \in (a,b) \cap D$ and

$$(10) \qquad P\{u_0 \leq \ell_s^{Y_s} < v\} \geq 1 - \varepsilon.$$

To construct $Y_s$ we will use the section theorem (see Dellacherie and Meyer [8, p. 137]) to define an increasing sequence of stopping times: $Y_s$ will be constant between terms of the sequence.

We can choose $S_0$ such that $T_0 \leq S_0 \leq T_0 + \frac{1}{2}\varepsilon$, $X_{S_0} \in D \cap (a,b)$: set $T_0' = \inf\{s > S_0: X_s \notin (a,b)\}$. Now use (9) to apply the section theorem to the set $\Gamma \cap (S_0, T_0' \wedge (T_0 + \varepsilon))$ to give a stopping time $S_1$ such that

$$P\{S_1 < \infty\} > 1 - \tfrac{1}{2}\varepsilon, \ \ell_{S_1}^{X_{S_1}} = u_0, \ X_{S_1} \in (a,b) \cap D,$$

and $S_1 < T_0' \wedge (T_0 + \varepsilon)$ on $\{S_1 < \infty\}$. Now put $T_1 = S_1' = \inf\{s > S_1: \ell_s^{X_{S_1}} = v_0\}$, and note that a.s. $T_1$ is a growth point of $\ell_s^{X_{S_1}}$ so that $X_{T_1} = X_{S_1} \in (a,b) \cap D$. Hence $T_1' = \inf\{s > T_1: X_s \notin (a,b)\} > T_1$, and if we put $T_1'' = \inf\{s > S_1: \ell_s^{X_{S_1}} = v\}$ we can again apply the section theorem to $\Gamma \cap (T_1, T_1' \wedge T'')$ to find a stopping time $S_2$ such that $P\{S_2 = \infty, S_1 < \infty\} < \frac{1}{4}\varepsilon$ and on $\{S_2 < \infty\}$ we have $T_1 < S_2 < T_1' \wedge T_1''$, $\ell_{S_2}^{X_{S_2}} = u_0$, and $X_{S_2} \in (a,b) \cap D$. Continuing inductively, we obtain, except on a set of probability $\varepsilon$, a sequence $(S_n)$, $(T_n)$ of stopping times such that $(T_n - S_n)$ are independent, identically distributed, $T_n < S_{n+1}$, $X_{S_n} \in (a,b) \cap D$ and $\ell_s^{X_{S_n}} \in [u_0, v)$ for $S_n \leq s < S_{n+1}$. If $Y_s = \sum 1_{[S_n, S_{n+1})}(s) X_{S_n}$ (10) is satisfied. But clearly $S_n = S_0 + (S_i - S_{i-1}) \geq S_0 + \sum (T_{i-1} - S_{i-1}) \to \infty$, so $Y_s$ is defined for all $s > T_0 + \varepsilon$ and the construction is valid outside a set of probability $\varepsilon$. Since $\varepsilon$ is arbitrary, this completes the proof.

We note that the conclusion of Proposition 4.2 allows us to deduce 'denseness' in two senses

COROLLARY 4.3. *Under the hypothesis of Proposition* 4.2, *for* $t > 0$, $0 \leq u < v \leq +\infty$,

$$\text{a.s. } \{x \in D: u < \ell_t^x < v\} \quad \text{is dense in } F_t.$$

COROLLARY 4.4. *Under the hypothesis of Proposition* 4.2, *for* $t > 0$, *if* $I$ *is an open interval with* $I \cap F_t \neq \phi$,

$$\text{a.s. } \{y: y = \ell_t^x \text{ for some } x \in D \cap I\} \text{ is dense in } \mathbb{R}^+.$$

REMARK. The totally asymmetric Cauchy process $A^{1,0,0,1}$ has a range which is a union of intervals, and therefore fails to satisfy (7). However the information in [26] can be used to show that its local time is dense in the sense of Proposition 4.2.

## 5. Constructions of $\ell_t^x$ that fail at some level

In the literature there are many distinct ways of obtaining $\ell_t^x$ as the limit of functionals $K_n(x,t)$ of the sample path $X_s$, $0 \leq s \leq t$. A systematic approach to these constructions was initiated by Maisonneuve [21], and developed into a unified umbrella method in Fristedt, Taylor [11], to which the reader is referred for a bibliography. Suppose that a construction $(K_n)$ converges a.s. at one level $x_0$, to give an additive functional $K_t^{x_0} = \lim_{n\to\infty} K_n(x_0,t)$, which is continuous in $t$, and which is normalised by (4), with $x=y=x_0$. A Fubini argument then shows that $K_n(x,\cdot)$ converges on a set of full measure in $\mathbb{R}_1$ and we can trivially use $(K_n)$ to define a version of the local time $\ell_t^x$ satisfying Theorem 2.1 by setting

$$\ell_t^x = \limsup_{n\to\infty} K_n(x,t).$$

Remembering from our earlier discussion that we only have a canonical value of $\ell_t^x$ for all $(x,t)$ whenever there is a continuous version, there are two distinct questions to resolve for any construction $K_n(x,t)$ which converges to $\ell_t^x$ $t\geq 0$ a.s. for each $x$.

I. *Is there a fixed null set* N *such that, for* $\omega \notin N$, $(K_n)$ *converges for all* $(x,t)$? *(If so, the result is automatically a version of* $\ell_t^x$ *satisfying Theorem* 2.1.)

II. *Suppose* $X_t$ *is such that a continuous version* $L_t^x$ *of the local time exists. Is there a null set* N *such that, for* $\omega \notin N$, $(K_n)$ *converges to* $L_t^x$ *for all* $(x,t)$?

We now give an example of a construction which fails to converge at some levels, thus showing that the answer to I may be "no." Let $N_t(x,x+\varepsilon)$ be the number of upcrossings made by F from $(-\infty,x)$ to $(x+\varepsilon,\infty)$ before time $t$. Suppose there exists a sequence $\varepsilon_k \downarrow 0$, and constants $a_k$ such that, for each $(x,t)$ a.s.,

$$a_k N_t(x,x+\varepsilon_k) \to \ell_t^x \quad \text{as } k\to\infty. \tag{11}$$

THEOREM 5.1. *Suppose* $X_t$ *is a Lévy process with local time* $\ell_t^x$, *and satisfying* (7), (8) *and* (11). *Then given* $0<u<v<+\infty$, $t>0$ *a.s., there exists a level* $z = z(w)$ *such that*

(12) $$\limsup_{k\to\infty} a_k N_t(z,z+\varepsilon_k) \geq v > u > \liminf_{k\to\infty} a_k N_t(z,z+\varepsilon_k).$$

PROOF: We use Proposition 4.2 to obtain $z$ as the limit point in a condensation argument. First note that, since D is countable, we can assume that a.s. (11) holds at every point of D. Apply Proposition 4.2 to find $y_0 \in D$ with $\ell_t^{y_0} > v$. But $y \to N_t(y,y+\varepsilon_{k_1})$ is constant in a small closed interval $[y_0', y_0'']$ with $y \in (y_0', y_0'')$. The Proposition now gives a point $y_1 \in (y_0', y_0'') \cap D$ for which $\ell_t^{y_1} < u$, and therefore for some $k_2 > k_1$, $a_{k_2} N_t(y_1, y_1+\varepsilon_{k_2}) < u$. By induction we obtain a sequence $I_r$ of closed intervals, which we may assume nested, such that

$$r \text{ even, } x \in I_r \Rightarrow a_{k_r} N_t(x,x+\varepsilon_{k_r}) < u$$

$$r \text{ odd, } x \in I_r \Rightarrow a_{k_r} N_t(x,x+\varepsilon_{k_r}) > v.$$

Clearly $z = \cap I_r$ satisfies (12).

REMARK 1. The asymmetric Cauchy process studied in Pruitt, Taylor [26] satisfies the conditions of the Theorem; the construction of its local time given there involved counting 'passes' of given length across a level, but for a fixed level this is equivalent to counting upcrossings.

REMARK 2. A similar argument, giving non-convergence at some level, will work for any construction $K_n(x,t)$ such that, a.s., $K_n(y,t) = K_n(x,t)$ for $y$ sufficiently close to $x$. For example, the analogue of Theorem 5.1 is valid for the Getoor-Millar construction (see [14]), which counts jumps across a level, and which we will consider in section 6.

The preceding counterexample deals with processes with a discontinuous local time: it might be thought that if $\ell_t^x$ is jointly continuous then any construction $(K_n)$ should converge simultaneously at every level $x$. In fact this is false even for $B_t$, as is shown in Barlow, Perkins [3]. We now give a generalization of their construction.

We start with a real variable result. Suppose $\psi: [0,1] \to \mathbb{R}$ is a fixed function, and define

$$(13) \qquad \Lambda_x(\psi) = \{t \in [0,1): \psi(t) = x \text{ and } \exists\ \delta > 0 \text{ with } \psi(s) \neq x \text{ for } t < s < t+\delta\};$$

denoting the starting points of excursions from $x$. Let $R_r = \{\psi(s): 0 \le s < r\}$ denote the range of $\psi$ with interior $R_r^0$ and closure $\bar{R}_r$.

THEOREM 5.2. *Suppose* $\psi:[0,1] \to \mathbb{R}$ *is cadlag, nowhere monotone and satisfies*

$$(14) \qquad R_r^0 \text{ is dense in } R_r \text{ for all } r \text{ in } (0,1].$$

*Let* $f:[0,1] \to [0,\infty)$ *be any continuous strictly increasing function with* $f(0) = 0$. *Then there is a set* $S$ *which is a countable intersection of sets each of which is open and dense in* $R_1^0$ *such that, for all* $x$ *in* $S$, *and* $t$ *in* $\Lambda_x(\psi)$, *there is a sequence* $\{t_n\}$ *decreasing to* $t$ *for which*

$$(15) \qquad |\psi(t_n) - \psi(t)| < f(t_n - t).$$

PROOF. For $0 < r \le 1$ and $x \in \bar{R}_r$ define

$$g_x^r = \sup\{s < r: \psi(s) = x \text{ or } \psi(s-) = x\}.$$

Then, for fixed $r$, $g_\cdot^r$ is upper semi-continuous on $\bar{R}_r$, that is,

$$g_x^r \ge \limsup_{\substack{y \to x \\ y \in \bar{R}_r}} g_y^r$$

However, for fixed $r$, we claim that the discontinuity points of $g^r_\cdot$ are dense in $\bar{R}_r$. For, suppose $(a,b) \cap \bar{R}_r \neq \phi$: using (14) we can assume without loss of generality that $\psi(r\pm) \notin [a,b]$ and $(a,b) \subseteq R^0_r$. If

$$M = \sup\{s < r: \psi(s) \in (a,b)\}$$

then $0 < M < r$, $\psi(M-) \in [a,b]$, and either $\psi(M-) > a$ or $\psi(M-) < b$: we assume $\psi(M-) > a$. The left continuity and nowhere monotonicity of $\psi(s-)$ imply there is a $t_1 < M$ such that

$$a < \inf\{\psi(s-): t_1 \leq s \leq M\} < \psi(t_1) \wedge \psi(t_1-).$$

Now choose the largest $t_0$ in $[t_1,M]$ satisfying

$$\psi(t_0) \wedge \psi(t_0-) = \inf\{\psi(s-): \ t_1 \leq s \leq M\} = x_0, \quad \text{say}.$$

Now $x_0 \in (a,\psi(M-)] \subset (a,b]$, For all $u \in (t_1,t_0) \cup (t_0,r)$

$$\psi(u), \psi(u-) \notin (a,x_0). \tag{16}$$

This is clear for $u \in (t_1,t_0)$ by the definition of $x_0$ and for $u \in (M,r)$ by the definition of $M$ and the fact that $(a,x_0) \subset (a,b)$. Finally, if $t_0 < M$, then $\psi(u) \wedge \psi(u-) > x_0$ for $t_0 < u \leq M$ by the choice of $t_0$. Now (16) implies

$$g^r_{x_0} = t_0 \quad \text{and} \quad \limsup_{y \uparrow x_0,\ y \in \bar{R}_r} g^r_y \leq t_1 < t_0.$$

(Note that the existence of $x_n \in \bar{R}_r$ with $x_n \uparrow x_0$ is guaranteed by $(a,b) \subset R^0_r$: this is all we need from (14).) Thus $x_0 \in (a,b]$ is a point of discontinuity of $g^r_x$, so the discontinuity points are dense.

Now let $f^{-1}$ be the continuous inverse of $f$, and define for

$r \in (0,1]$ and $n \in \mathbb{N}$,

$$G_n^r = (R_1^0 \backslash \bar{R}_r) \cup \{x \in R_r^0 : \exists\, \varepsilon > 0 \text{ with } (x-\varepsilon, x+\varepsilon) \subset R_r^0 \text{ and for all}$$

$$y \in (x-\varepsilon, x+\varepsilon), \exists\, h = h(y) \in (-n^{-1}, n^{-1}) \text{ with } g_{y+h}^r - g_y^r > f^{-1}(|h|)\}.$$

Clearly, $G_n^r$ is open. We now show it is dense in $R_1^0$. Suppose $(a,b) \subset R_r^0$: then $(a,b)$ contains a discontinuity of $g_\cdot^r$. Pick $\varepsilon_0 > 0$ and $x, x' \in (a,b)$ such that

$$2\varepsilon_0 < g_{x'}^r - g_x^r, \; |x - x'| < n^{-1} \text{ and } f^{-1}(|x-x'|) < \varepsilon_0.$$

Use the continuity of $f^{-1}$ and upper semi-continuity of $g^r$ to find $\varepsilon < n^{-1} - |x-x'|$, $\varepsilon > 0$ and such that $f^{-1}(\varepsilon + |x-x'|) < \varepsilon_0$ and $y \in (x-\varepsilon, x+\varepsilon)$ implies $y \in R_r^0$ and $g_y^r \le g_x^r + \varepsilon_0$. For such $y$ take $h(y) = x - y + (x'-x) \in (-n^{-1}, n^{-1})$ and check that $g_{y+h}^r - g_y^r = g_{x'}^r - g_y^r > \varepsilon_0 > f^{-1}(\varepsilon + |x-x'|) \ge f^{-1}(|h|)$. Thus $x \in G_n^r \cap (a,b)$ and it follows that $G_n^r$ is dense in $R_1^0$. Since $R_1^0$ is locally compact, the Baire category theorem implies that

$$S = \bigcap_{\substack{r \in Q \cap (0,1] \\ n \in \mathbb{N}}} G_n^r \quad \text{is a dense } G_\delta \text{ in } R_1^0.$$

Now let $x \in S$ and $t \in \Lambda_x(\psi)$. Then there are $r_n \in Q$ such that $r_n \downarrow t$ and $t = g_x^{r_n}$. But $x \in S \cap \bar{R}_{r_n}$, so we can find a sequence $h_n$ with $h_n \to 0$ such that

$$g_{x+h_n}^{r_n} - g_x^{r_n} > f^{-1}(|h_n|).$$

Since $t_n = g_{x+h_n}^{r_n} \in (t, r_n]$, we have

$$|\psi(t_n) - \psi(t)| \wedge |\psi(t_{n-}) - \psi(t)| \le |x + h_n - x| < f(t_n - t).$$

By slightly moving the $t_n$'s, if necessary, we get

$$|\psi(t_n) - \psi(t)| < f(t_n - t), \quad t_n \downarrow t,$$

and the proof is complete.

REMARK 1. The theorem is proved in [3] under the additional hypothesis that $\psi(s)$ is continuous.

REMARK 2. The theorem is false if (14) is omitted. Suppose $\psi(\cdot)$ is cadlag, nowhere monotone, such that for each $x$ there is at most one value of $t$ for which $x = \psi(t)$ or $x = \psi(t-)$. One could easily construct such a function directly, but note that any symmetric stable process of index $\alpha \le \frac{1}{2}$ has sample paths which a.s. have this property (see [28]). If we define $g_{\cdot}^{r}$ as in the proof of the Theorem, one can check that it is continuous, and hence uniformly continuous on $\bar{R}_1$. Therefore there is a continuous strictly increasing function $\phi$ such that

$$x,y \in \bar{R}_1 \Rightarrow |g_x' - g_y'| < \phi(|x-y|).$$

If $f = \phi^{-1}$, it follows that

$$|\psi(t) - \psi(s)| > f(|t-s|) \quad \text{for all} \quad s,t \in [0,1)$$

and the theorem fails for $f$.

COROLLARY 5.3. *Suppose* X *is a Lévy process which a.s. has a continuous local time, then the conclusion of Theorem* 5.2 *a.s. holds for* $\psi(s) = X_s$.

PROOF: We need only check that $X_s$ a.s. satisfies the hypothesis of the Theorem. The existence of a continuous local time $L_t^x$ implies that the sample path $X_s$ is nowhere monotone by a real variable argument (see Example 1 in Geman, Horowitz [12]). To prove (14) note that a.s.

$$\int_0^t I(X_s \in B)ds = \int_B L_t^x dx \quad \text{for all Borel } B,\ t \geq 0.$$

It follows that $\{x: L_r^x > 0\} \subset R_r^0$ and is dense in $R_r$ for all $r > 0$.

EXAMPLE 5.4. Let $X$ be a fixed Lévy process with a jointly continuous local time $L_t^x$. If $V$ denotes an excursion of $X$ from $x$, we put $\tau^-(V)$ and $\tau^+(V)$ for the start and end of $V$. There exists a continuous, strictly increasing function $f$ with $f(0) = 0$ which grows slowly enough to ensure

$$\liminf_{h \downarrow 0} |X_{\tau^-(V)+h} - x|/f(h) \geq 1 \tag{17}$$

holds a.s. for fixed $X, V$ and so for fixed $x$ it holds a.s. for every excursion $V$ from $x$. To find such an $f$, let $\mu$ denote the characteristic measure of the Poisson point process of excursions $V$ from $x$ (see Itô [16] or Fristedt, Taylor [11] for details), fix $u \in (0,1)$, and choose $\varepsilon_n \downarrow 0$ such that

$$\mu\{V: \inf\{|V_s|;\ s \in [u^{n+1}, u^n] < \varepsilon_n \ \text{ and } \ l(V) > u^n\} < 2^{-n}.$$

Here $\ell(V) = \tau^+(V) - \tau^-(V)$ is the length of the excursion and $V_s = X_{s-\tau^-(V)}$ for $0 \leq s < l(V)$. Define $f(u^n) = \varepsilon_n$ and by linear interpolation in $(u^{n+1}, u^n)$, $n \in \mathbb{N}$. Then a standard Borel-Cantelli argument shows that

$$\mu\{V: \inf(V_s/f(s): u^{n+1} \leq s \leq u^n) < 1 \text{ i.o.}\} = 0.$$

This establishes (17) for this function f.

Let $N_\varepsilon(t,x)$ be the number of excursions from x exceeding $\varepsilon$ in length and completed by time t, and let $N'_\varepsilon(t,x)$ be the number of these excursions that satisfy (17). For fixed x, $N_\varepsilon(t,x) = N'_\varepsilon(t,x)$ a.s.

By Maisonneuve [21, Theorem X.4], if $E_\varepsilon$ denotes the subset of excursion space consisting of excursions of length greater than $\varepsilon$, we have, for each $x \in R$,

$$(18) \qquad \lim_{\varepsilon \downarrow 0} N'_\varepsilon(t,x)\mu(E_\varepsilon)^{-1} = L_t^x \quad t \geq 0 \ \text{a.s.}$$

But Corollary 5.3 tells us that a.s. there is a dense $G_\delta$ in $R_1^0$, $S = S(\omega)$ such that, for $x \in S$ and $t \in \Lambda_x(X)$,

$$\liminf_{h \downarrow 0} |X_{t+h} - X_t|/f(h) = 0.$$

This shows that, for $x \in S$, (17) fails for every excursion which starts from x. There are only countably many levels x at which some excursion from x begins with a jump. Thus, if $x \in S'(\omega) = S(\omega)\setminus\{X(t-): X(t) \neq X(t-)\}$, (17) fails for every excursion starting from x, so that $N'_\varepsilon(t,x) = 0$ for all $0 \leq t \leq 1$. Thus we have found a random set $S'(\omega)$ that is a dense $G_\delta$ in $R_1^0$, and for which (18) fails for $x \in S'(\omega)$.

Note that if $t_0(x)$ is the first time $t \in \Lambda_x(B)$ for which (17) fails with $f(h) = h$ and $Q_t(\omega) = \{x: t_0(x) \leq t\}$, then $Q_t$ is of the form considered in (6).

EXAMPLE 5.5. The construction above gives a counterexample to Question II, but it is not obvious that $N'_\varepsilon(t,x)\mu(E_\varepsilon)^{-1}$ fails to converge for some levels x. We now give such an example. Suppose the process is Brownian motion: then Perkins [23] showed that outside a fixed null set $N$,

$$\lim_{\varepsilon \downarrow 0} (\pi/2)^{\frac{1}{2}} N_\varepsilon(t,x)\varepsilon^{\frac{1}{2}} = L_t^x \quad \text{for all} \quad (x,t). \tag{19}$$

We now define $N''_\varepsilon(t,x)$ by counting all excursions from $x$ completed by time $t$ with $\ell(V) > \varepsilon$, and for which either $2^{-2k} \le \ell(V) < 2^{-2k+1}$ for some $k \in N$, or $2^{-2k-1} \le \ell(V) < 2^{-2k}$ for some $k \in \mathbb{N}$ and (17) holds. If we now look at any point $x \in S \subset R_t$, (19) shows that $\varepsilon^{\frac{1}{2}} N''(t,x)$ cannot converge as $\varepsilon \downarrow 0$, for $0 \le t \le 1$. Clearly $N''_\varepsilon(t,x) = N_\varepsilon(t,x)$ a.s. for a fixed level $x$, so we have a construction for Brownian local time which fails to converge at some levels, giving a counterexample to Question I.

Both the constructions above depend on the behaviour of $X_s$ away from the level $x$. A construction of $L_t^x$ which depends only on the level set $\{s \le t: X_s = x\}$ is called intrinsic. One could ask whether negative answers to I and II are possible for an intrinsic construction. We will obtain such a counterexample again based on Brownian local time.

EXAMPLE 5.6. Let $f(t) = e^{-t^2}$, $t > 0$; consider the set of ending points of Brownian excursions from $x$

$$\Gamma_x(B) = \{t > 0: B_t = x,\ x \ne B_{t-h} \quad \text{for} \quad h \in (0,\delta), \quad \text{some} \quad \delta > 0\}.$$

Since the points $t$ in $\Gamma_x(B)$ are stopping times we can apply the usual integral test for the lower asymptotic growth rate of $L_{t+h}^x$ for small $h > 0$ to see that, for fixed $t \in \Gamma_x(B)$.

$$(L_{t+h}^x - L_t^x) f(h)^{-1} \to \infty \quad \text{as} \quad h \downarrow 0 \quad \text{a.s.} \tag{20}$$

and therefore, for each fixed $x$, a.s. (20) is true for all excursions from x. Hence if we put $N_\varepsilon'''(t,x)$ for the number of excursions from x which satisfy (20), then for fixed x,

$$\lim_{\varepsilon\downarrow 0} (\pi/2)^{\frac{1}{2}}\varepsilon^{\frac{1}{2}}N_\varepsilon'''(t,x) = L_t^x \quad \text{for all} \quad t \geq 0, \text{ a.s.}$$

The condition (20) is intrinsic to the level set at x because of the uniform result (19). However we claim that there is a dense $G_\delta$ set $S_1 = S_1(\omega)$ such that for all $x \in S_1$, $t \in \Gamma_x(B)$ we have

$$\liminf_{h\downarrow 0} (L_{t+h}^x - L_t^x)f(h)^{-1} \leq 1; \tag{21}$$

so that for such levels x, $N_\varepsilon'''(t,x) = 0$ for all t.

For $r \geq 0$, $x \in \mathbb{R}$, let

$$T_r(x) = \inf\{t \geq r\colon B_t = x\},$$

$$G_n^r = \{x\colon \varepsilon > 0,\ h \in (0,n^{-1}) \text{ such that}$$

$$y \in (x-\varepsilon,x+\varepsilon) \Rightarrow L_{T_r(y)+h}^y - L_{T_r(y)}^y < f(h)\}.$$

If $\omega$ is chosen so that $B_t(\omega)$ is nowhere monotone and $L_t^x(\omega)$ is continuous, it is clear that $G_n^r$ is open. We now show it is dense. For any open interval (a,b) let us assume $T_r(a) < T_r(b)$. As B is nowhere monotone we can find $t \in (T_r(a),T_r(b))$ such that

$$B_t < \sup(B_s\colon r \leq s \leq t) \equiv x_0 \in (a,b),$$

and hence a $\delta > 0$ such that $B_{T_r(x_0)+h} < x_0$ for $0 < h < \delta$. Fix $h < \delta \wedge n^{-1}$ and note that $\lim_{y\uparrow x_0} T_r(y) = T_r(x_0)$, so that

$$\lim_{y\uparrow x_0} L_{T_r(y)+h}^y - L_{T_r(y)}^y = L_{T_r(x_0)+h}^{x_0} - L_{T_r(x_0)}^{x_0} = 0.$$

It follows that, for some $\varepsilon > 0$, $(x_0-\varepsilon,x_0) \subset G_n^r$ and hence $G_n^r \cap (a,b) \neq \phi$.

We now take

$$S_1(\omega) = \bigcap_{\substack{n\in\mathbb{N} \\ r\in Q,\ r\geq 0}} G_n^r$$

and deduce that $S_1(\omega)$ is a dense $G_\delta$ by the Baire category theorem. Since for each $t \in \Gamma_x(B)$ we have $t = T_r(x)$ for some rational $r$, we have proved that for every $x \in S_1$, $t \in \Gamma_x(B)$ we have (21). This establishes the claim and completes the example.

## 6. Some constructions for $L_t^x$ which converge at all levels

The following theorem extends a non-intrinsic construction of Getoor, Millar [14] to all levels simultaneously.

THEOREM 6.1. *Suppose* $X_t$ *is a Lévy process with measure* $\nu$ *as defined in* (1) *satisfying* $\int(|x|\wedge 1)\nu(dx)=\infty$, *and assume* X *has a jointly continuous local time* $L_t^x$. *Define*

$$f_\varepsilon^a(x,y) = I_{\{x<a-\varepsilon,\ y>a+\varepsilon\}} + I_{\{x>a+\varepsilon,\ y<a-\varepsilon\}},$$

$$Q_\varepsilon^a(t) = \sum_{0\leq s\leq t} f_\varepsilon^a(X_{s-},X_s),$$

$$b_\varepsilon = \int_{-1}^{1} (|x| - 2\varepsilon)\vee 0\ \nu(dx).$$

*If there exist sequences* $\varepsilon_n \downarrow 0$, $\delta_n \downarrow 0$ *with* $0<\delta_n<\varepsilon_n$ *such that, as* $n\to\infty$

(22) $$\frac{b_{\varepsilon_n-\delta_n}}{b_{\varepsilon_n+\delta_n}} \to 1,$$

(23) $$b_{\varepsilon_n} / |\log \delta_n| \to +\infty,$$

*then, for each* $\eta, t_0 > 0$,

$$\lim_{n\to\infty} P\left(\sup_{a\in\mathbb{R},\, s\le t_0} |b_{\varepsilon_n}^{-1} Q_{\varepsilon_n}^{a}(s) - L_s^a| > \eta\right) = 0.$$

*If, in addition, as* $n \to \infty$

(24) $$b_{\varepsilon_n} \cdot b_{\varepsilon_{n+1}}^{-1} \to 1,$$

(25) $$\sum_n e^{-\theta b_{\varepsilon_n}} \text{ converges for all } \theta > 0,$$

*then, for each* $t_0$, a.s. *as* $\varepsilon \downarrow 0$

$$b_\varepsilon^{-1} Q_\varepsilon^a(s) \to L_s^a \text{ uniformly for } a \in \mathbb{R},\ 0 \le s \le t_0.$$

REMARK. We know of no Lévy process with a continuous local time and satisfying $\int(|x| \wedge 1)\nu(dx) = \infty$, that fails to satisfy all the hypotheses of the above theorem.

PROOF: Let $L_t^* = \sup\{L_t^x,\ x \in \mathbb{R}\}$, and $T = \inf\{t\colon L_t^* = 1\}$. It is clearly sufficient to prove the theorem with $T$ in place of $t_0$.

We introduce the notation

$$N(x,dy) = \nu(dy - x)$$

$$Nf_\varepsilon^a(x) = \int N(x,dy)\, f_\varepsilon^a(x,y) = \begin{cases} \nu[a+\varepsilon-x,\infty), & \text{if } x \le a-\varepsilon \\ \nu(-\infty,a-\varepsilon-x], & \text{if } x \ge a+\varepsilon \end{cases}$$

$$V_\varepsilon^a(t) = \int_0^t Nf_\varepsilon^a(X_{s-})ds = \int_{-\infty}^{\infty} Nf_\varepsilon^a(y) L_t^y dy, \text{ by (5).}$$

We will first use the continuity of $L_t^x$ to show

(26) $$b_\varepsilon^{-1} V_\varepsilon^a(t) \to L_t^a \text{ uniformly in } (t,a) \in [0,T]\times\mathbb{R} \text{ as } \varepsilon \to 0 \text{ a.s.,}$$

and then consider the difference

$$M_\varepsilon^a(t) = Q_\varepsilon^a(t) - V_\varepsilon^a(t).$$

By considering the process

$$X_t' = X_t - \sum_{0\le s\le t} (X_s - X_{s-})\, I_{\{|X_s - X_{s-}| \ge 1\}}$$

and working between consecutive jumps of $X$ of magnitude exceeding 1, we may assume without loss of generality that

$$\nu(-\infty,-1] = \nu[1,\infty) = 0.$$

Now

(27) $$\int_{-\infty}^{\infty} Nf_\varepsilon^0(x)dx = \int_{-\infty}^{a-\varepsilon} \nu[a+\varepsilon-x,\infty)dx + \int_{a+\varepsilon}^{\infty} \nu(-\infty,a-\varepsilon-x]dx$$

$$= \int_{-\infty}^{\infty} (|u| - 2\varepsilon) \vee 0 \;\nu(du) \;=\; b_\varepsilon,$$

where we have used support $(\nu) \subset [-1,1]$ in the last line. As $\int(|x| \wedge 1)\nu(dx) = \infty$ and $\int(x^2 \wedge 1)\nu(dx) < \infty$, $b_\varepsilon$ is finite for all $\varepsilon > 0$ and $b_\varepsilon \uparrow \infty$ as $\varepsilon \downarrow 0$. We now write

(28) $$b_\varepsilon^{-1} V_\varepsilon^a(t) - L_t^a = b_\varepsilon^{-1} \int_{-\infty}^{\infty} Nf_\varepsilon^a(y)(L_t^y - L_t^a)dy.$$

For any $\eta > 0$ we can first choose $\delta > 0$ such that

$$\sup\{|L_t^x - L_t^y| : |x-y| \le \delta,\ t \le T\} < \eta,$$

since $L_t^x$ is uniformly continuous in $(x,t) \in \mathbb{R} \times [0,T]$. Also $\int_{|y-a|<\delta} Nf_\varepsilon^a(y)dy \le \int_{-\infty}^{\infty} Nf_\delta^a(x)dx = b_\delta$, whenever $\varepsilon < \delta$, so we can split the integral in (28) into two parts, $|y-a| \le \delta$ and $|y-a| > \delta$, to give

$$\sup_{t\le T}|b_\varepsilon^{-1}\ V_\varepsilon^a(t) - L_t^a| \le \eta + b_\varepsilon^{-1} b_\delta L_T^* .$$

Now let $\varepsilon \downarrow 0$, and we have a uniform bound which establishes (26).

We now consider $M_\varepsilon^a(t)$. As $(N(x,dy),t)$ is a Lévy system for $X$, $M_\varepsilon^a(t)$ is a martingale (see Benveniste-Jacod [5]) with $\langle M_\varepsilon^a, M_\varepsilon^a\rangle_t = V_\varepsilon^a(t)$.

LEMMA 6.2. *For all* $y \in (0,1)$, $\varepsilon > 0$, *and* $a \in \mathbb{R}$,

$$P(\sup_{t\ge 0} b_\varepsilon^{-1}|M_\varepsilon^a(t\wedge T)| > y) \le 2e^{-\frac{1}{4}b_\varepsilon y^2} .$$

PROOF: The definition of $T$ and (27) imply $\langle M_\varepsilon^a, M_\varepsilon^a\rangle_T \le b_\varepsilon$. If the supremum is taken over $t \in \{i/n: i;0,\dots,n^2\}$, the result is an easy consequence of Theorem 1.6 of Freedman [9]. Now let $n \to \infty$.

Fix $K > 0$ and let

$$A_n = \{i\delta_n,\ i \in \mathbb{Z}\} \cap [-K,K].$$

If $y \in (0,1)$, the above lemma gives

$$\text{(29)}\qquad P(\sup\{b_{\varepsilon_n}^{-1}|M_{\varepsilon_n}^a(t)|,\ a \in A_n,\ 0\le t\le T\} > y)$$
$$\le 2\ \exp\left\{\log\left(\frac{2K}{\delta_n} + 1\right) - \tfrac{1}{4}b_{\varepsilon_n} y^2\right\}$$

which converges to zero by (23). If $x \in [-K,K]$ and $a_n(x)$ is the (suitably defined) "nearest" point to $x$ in $A_n$, then

$$Q_{\varepsilon_n+\delta_n}^{a_n(x)}(t) \le Q_{\varepsilon_n}^x(t) \le Q_{\varepsilon_n-\delta_n}^{a_n(x)}(t).$$

Hence

$$\text{(30)}\quad |b_{\varepsilon_n}^{-1}Q_{\varepsilon_n}^x(t) - L_t^x| \le |b_{\varepsilon_n}^{-1}Q_{\varepsilon_n-\delta_n}^{a_n(x)}(t) - L_t^x| + |b_{\varepsilon_n}^{-1}Q_{\varepsilon_n+\delta_n}^{a_n(x)}(t) - L_t^x|.$$

Each of these two terms may be bounded in a similar fashion; taking the first we have

$$|b^{-1}_{\varepsilon_n}Q^{a_n(x)}_{\varepsilon_n-\delta_n}(t) - L^x_t| \le b^{-1}_{\varepsilon_n}|Q^{a_n(x)}_{\varepsilon_n-\delta_n}(t) - V^{a_n(x)}_{\varepsilon_n-\delta_n}(t)|$$

$$+ b_{\varepsilon_n-\delta_n} b^{-1}_{\varepsilon_n}|b^{-1}_{\varepsilon_n-\delta_n} V^{a_n(x)}_{\varepsilon_n}(t) - L^{a_n(x)}_t|$$

$$+ |b_{\varepsilon_n-\delta_n} b^{-1}_{\varepsilon_n} L^{a_n(x)}_t - L^x_t| .$$

Using (29), (22), (23) and (26), and the joint continuity of $L^x_t$ we deduce that, for each $y \in (0,1)$

$$\lim_{n\to\infty} P(\sup\{|b^{-1}_{\varepsilon_n}Q^x_{\varepsilon_n}(t) - L^x_t| : x \in [-K,K],\ 0\le t\le T\} > y) = 0,$$

proving the first assertion in the theorem.

Now condition (25), applied to (29) gives, for each $y \in (0,1)$, a convergent series. An application of Borel Cantelli now gives

$$\sup\{b^{-1}_{\varepsilon_n}|M^x_{\varepsilon_n}(s)| : x \in A_n,\ 0\le s\le T\} \to 0 \quad \text{a.s.}$$

and hence by (30) and the argument following

$$\sup\{|b^{-1}_{\varepsilon_n}Q^x_{\varepsilon_n}(t) - L^x_t| : x \in [-K,K],\ 0\le s\le T\} \to 0 \quad \text{a.s.}$$

However, if $\varepsilon_{n+1} \le \varepsilon \le \varepsilon_n$,

$$\left(b_{\varepsilon_n} b^{-1}_{\varepsilon_{n+1}}\right) b^{-1}_{\varepsilon_n}Q^x_{\varepsilon_n} \le b^{-1}_{\varepsilon} Q^x_{\varepsilon} \le b_{\varepsilon_{n+1}} Q^x_{\varepsilon_{n+1}} \left(b_{\varepsilon_{n+1}} b^{-1}_{\varepsilon_n}\right)$$

and therefore, by (24) we have a.s.

$$b_\varepsilon^{-1} Q_\varepsilon^x(t) \to L_t^x \quad \text{uniformly in} \quad x \in [-K,K],\ t \in [0,T]. \qquad \square$$

EXAMPLE 6.3. Symmetric stable process of index $\alpha$ , $1 < \alpha < 2$ has Lévy measure $\nu(dx) = x^{-1-\alpha}\,dx$ so that $b_\varepsilon \sim c\varepsilon^{1-\alpha}$ as $\varepsilon \downarrow 0$. Take $\varepsilon_n = \frac{1}{n}$, $\delta_n = \frac{1}{n^2}$ to satisfy conditions (22) to (25). Boylan [6] proved that $S^{\gamma,0}$ has a continuous local time so the conclusion of our theorem is valid.

EXAMPLE 6.4. Critical asymmetric process $A^{1,\alpha,0,p}$ has a continuous local time for $\alpha > 0$. In this case

$$b_\varepsilon \sim c(\log \frac{1}{\varepsilon})^{\alpha+1} \quad \text{as} \quad \varepsilon \downarrow 0.$$

Take $\varepsilon_n = e^{-n}$, $\delta_n = e^{-n-1}$ and all the conditions (22) to (25) are satisfied for $\alpha > 0$. Again we a.s. get uniform convergence for the construction.

EXAMPLE 6.5. Critical symmetric process $S^{1,\alpha,0}$ satisfies (22) to (25) if $\alpha > 0$, as in the asymmetric case. If $\alpha > 2$, $L_t^x(S^{1,\alpha,0})$ is jointly continuous and the theorem applies. If $\alpha \in (0,2)$, $L_t^x(S^{1,\alpha,0})$ is discontinuous. Nevertheless, the $L^1$-continuity of local time (in space) may be used in (28) to show $b_\varepsilon^{-1} V_\varepsilon^a(t) \xrightarrow{L^1} L_t^a$ as $\varepsilon \to 0$ for each a, t, and the rest of the proof goes through to show

$$\lim_{\varepsilon\to 0} \sup_{a,t\le t_0} P(|b_\varepsilon^{-1} Q_\varepsilon^a(t) - L_t^a| > \eta) = 0 \quad \text{for} \quad \eta > 0,\ t_0 > 0.$$

Choose $\varepsilon_n \downarrow 0$ such that $b_{\varepsilon_n}^{-1} Q_{\varepsilon_n}^a(t) \to L_t^a$ a.s. for each a, t. If $\alpha = 1$ and $\beta \in (1,2)$ then $F_t$ is nowhere dense by Kesten [18] and $L_t^x(S^{1,\alpha,0})$ is unbounded on $D \cap (-\eta,\eta)$ $(t,\eta > 0)$ by Millar and Tran [22]. By Theorem 5.1 (and the subsequent Remark 2), a.s. there exists $a = a(\omega)$

such that $\{b_{\varepsilon_n}^{-1} Q_{\varepsilon_n}^a(t)\}$ fails to converge as $n \to \infty$.

We now consider two "intrinsic" constructions of local time.

The first is the characterization of $\ell_t^x$ as the appropriate Hausdorff measure of the level set

$$Z_X(x,t) = \{s \in [0,t]: X_s = x\}.$$

This construction, which is described in Taylor [29], depends on the fact that for fixed x, the inverse of $\ell_\cdot^x$ is a subordinator, with each jump corresponding to an excursion of X from the level x. Fristedt, Pruitt [10] showed that for each subordinator there is a Hausdorff measure function which makes the measure of the range up to the time t grow linearly with t. Using the method introduced by Taylor, Wendel [30] for B, $S^{\gamma,\delta}$, $1<\gamma<2$, they concluded that for any Lévy process, X, with a local time there is a $\phi$ such that

$$(31) \qquad \phi - m(Z_X(x,t)) = \ell_t^x \quad t \geq 0 \text{ a.s. for each } x.$$

Here $\phi - m$ denotes Hausdorff $\phi$-measure. Note that the "convergence" of this construction is built into the definition of $\phi$-measure, so that there is trivially an affirmative answer to Question I of section 5. Turning now to Question II we have the following result, that will be proved in [3]:

THEOREM 6.6. (a) *If* $\phi(s) = (2s \operatorname{loglog} \frac{1}{s})^{\frac{1}{2}}$, $s < e^{-1}$, *then*

$$\phi - m(Z_B(x,t)) = L_t^x(B) \quad \text{for all } (x,t) \text{ a.s.}$$

(b) *If* $\phi_\alpha(s) = \rho^{-\gamma}\gamma^{-\gamma}(1-\gamma)^{\gamma-1}s^\gamma(\operatorname{loglog}\frac{1}{s})^{1-\gamma}$, $s < e^{-1}$, *where* $\gamma = 1 - \frac{1}{\alpha}$ *and* $\rho^{-\gamma}$ *is given by* (1) *of* [15], *then for* $1 < \alpha < 2$,

$$\phi_\alpha - m(Z_{S^{\alpha,\delta}}(x,t)) = L_t^x(S^{\alpha,\delta}) \quad \text{for all } (x,t) \text{ a.s.}$$

(c) *If* $\psi_\alpha(s) = n^{-2}(\alpha-1)^{-1}(\log\frac{1}{s})^{1-\alpha}\log(\log(\log\frac{1}{s}))$, $s < e^{-e}$, *then for* $\alpha > 2+\sqrt{2}$,

$$\psi_\alpha - m(Z_{S^{1,\alpha,0}}(x,t)) = L_t^x(S^{1,\alpha,0}) \quad \textit{for all} \quad (x,t) \quad \textit{a.s.}$$

REMARKS. (a) was proved by Perkins [24], but the argument used the Ray-Knight theorems on Brownian local time and hence does not extend to other Lévy processes.

(c) indicates that the scaling properties of the stable processes are not needed in the proofs. Indeed we feel confident that our methods will apply to a large class of Lévy processes. However, as the conditions on $\alpha$ are stronger than those needed to ensure continuity of $L(S^{1,\alpha,0})$, we are clearly unable to prove the theorem for every Lévy process with a continuous local time.

CONJECTURE 6.6. *If* X *is a Lévy process with a continuous local time* $L_t^x$, *and* $\phi$ *is the Hausdorff measure function giving* (31), *then*

$$\phi - m(Z_X(x,t)) = L_t^x \quad \textit{for all} \quad (x,t) \quad \textit{a.s.}$$

There is another interesting intrinsic construction due to Kingman [19]. He showed that for any $X_t$ with a local time, there is a suitable monotone function $\psi(h)$ such that, for fixed $x$, a.s.

$$\psi(\varepsilon)|Z(x,t)(\varepsilon)| \to \ell_t^x \quad \text{for all} \quad t > 0, \tag{32}$$

where $|E|$ denotes the Lebesgue measure of $E$, and $E(\varepsilon) = \{s\colon \exists\, u \in E$ with $|u-s| < \frac{\varepsilon}{2}\}$. There is no a priori reason why (32) should converge simultaneously at all levels, but again we can obtain a positive result if we keep away from the critical cases.

THEOREM 6.7. (a) $\lim_{\varepsilon \to 0^+} \sup_{X \in \mathbb{R}, 0 \le t \le T} |\varepsilon^{-\frac{1}{2}} |Z_B(x,t)(\varepsilon)| - L_t^x(B)| = 0$ $\forall$ T > 0 a.s.

(b) *for each* $\alpha \in (1,2)$ *there is a* $c_{\alpha,\delta}$ *such that*

$$\lim_{\varepsilon \to 0^+} \sup_{x \in \mathbb{R}, 0 \le t \le T} |c_{\alpha,\delta} \cdot \varepsilon^{-1/\alpha} |Z_{S^{\alpha,\delta}}(x,t)(\varepsilon)| - L_t^x(S^{\alpha,\delta})| = 0$$

*for all* $T > 0$ *a.s.*

(c) *For each* $\alpha > 3$, *there is a* $c_\alpha$ *such that*

$$\lim_{\varepsilon \to 0^+} \sup_{x \in \mathbb{R}, 0 \le t \le T} |c_\alpha (\log \tfrac{1}{\varepsilon})^{1-\alpha} \varepsilon^{-1} |Z_{S^{1,\alpha,0}}(x,t)(\varepsilon)| - L_t^x(S^{1,\alpha,0})| = 0$$

*for all* $T > 0$ *a.s.*

REMARK. (a) is proved in Perkins [23]. The proof of (b) and (c) is given in [3], where the interested reader may find the values of $c_{\alpha,\delta}$ and $c_\alpha$.

## References


1. M.T. BARLOW. Zero-one laws for the excursions and range of a Lévy process. *Z. Wahrscheinlichkeitstheorie verw. Gebiete 55* (1981), 149-163.

2. M.T. BARLOW. Continuity of local times for Lévy processes. To appear.

3. M.T. BARLOW, E. PERKINS. Levels at which every Brownian excursion is exceptional. *Séminaire de Probabilités* XVIII, 1-28 (1984).

4. M.T. BARLOW, E. PERKINS, S.J. TAYLOR. Two uniform intrinsic constructions for the local time of a class of Lévy processes. To appear in *Ill. J. Math.*

5. A. BENVENISTE, J. JACOD. Systèmes de Lévy de processus de Markov. *Invent. Math. 21*, (1973), 183-198.

6. E.S. BOYLAN. Local times for a class of Markov processes. *Ill. J. Math. 8*, (1964), 19-39.

7. J. BRETAGNOLLE. Résultats de Kesten sur les processes à acroissements independents. *Séminaire de Probabilités V*, (1971), 21-36.

8. C. DELLACHERIE, P.A. MEYER. *Probabilités et Potentiel.* Vol. I, Hermann, Paris, 1975.

9. D. FREEDMAN. On tail probabilities for martingales. *Annals of Prob. 3*, (1975), 100-118.

10. B.E. FRISTEDT, W.E. PRUITT. Lower functions for increasing random walks and subordinators. *Z. Wahrscheinlichkeitstheorie verw. Gebiete 18*, (1971), 167-182.

11. B.E. FRISTEDT, S.J. TAYLOR. Constructions of local time for a Markov process. *Z. Wahrscheinlichkeitstheorie verw. Gebiete 62*, (1983), 73-112.

12. D. GEMAN, J. HOROWITZ. Occupation densities. *Annals Prob. 8*, (1980), 1-67.

13. R.K. GETOOR, H. KESTEN. Continuity of local times for Markov processes. *Compos. Math. 24*, (1972), 277-303.

14. R.K. GETOOR, P.W. MILLAR. Some limit theorems for local time. *Compos. Math. 25*, (1972), 123-134.

15. J. HAWKES. A lower Lipschitz condition for the stable subordinator. *Z. Wahrscheinlichkeitstheorie verw. Gebiete 17*, (1971), 23-32.

16. K. ITÔ. Poisson point processes attached to Markov processes. *Proc. of Sixth Berkeley Symposium*, (1970), 225-239.

17. H. KESTEN. Hitting probabilities of single points for processes with stationary independent increments. *Memoir 93 AMS* (1969).

18. H. KESTEN. Lévy processes with a nowhere dense range. *Indiana Univ. Math. J. 25*, (1976), 45-64.

19. J.F.C. KINGMAN. An intrinsic description of local time. *J. London Math. Soc. 6*, (1973), 725-731.

20. F. KNIGHT. Random walks and the sojourn density process of Brownian motion. *Trans. Amer. Math. Soc. 107*, (1963), 56-86.

21. B. MAISONNEUVE. Systèmes régéneratifs. *Astérisque 15*, (1974).

22. P.W. MILLAR, L.T. TRAN. Unbounded local times. *Z. Wahrscheinlichkeitstheorie verw. Gebiete 30*, (1974), 87-92.

23. E. PERKINS. A global intrinsic characterisation of Brownian local time. *Annals Prob. 9*, (1981), 800-817.

24. E. PERKINS. The exact Hausdorff measure of the level sets of Brownian motion. *Z. Wahrscheinlichkeitstheorie verw. Gebiete 58*, (1981), 373-388.

25. E. PERKINS. On the continuity of the local time of stable processes. To appear.

26. W.E. PRUITT, S.J. TAYLOR. The local structure of the sample paths of asymmetric Cauchy processes. To appear.

27. D.B. RAY. Sojourn times of a diffusion process. *Ill. J. Math. 1*, (1963), 615-630.

28. S.J. TAYLOR. Multiple points for the sample paths of the symmetric stable process. *Z. Wahrscheinlichkeitstheorie verw. Gebiete 5*, (1966), 247-264.

29. S.J. TAYLOR. Sample path properties of processes with stationary independent increments, in *Stochastic Analysis*, 387-414. Wiley, London, 1973.

30. S.J. TAYLOR, J.G. WENDEL. The exact Hausdorff measure of the zero set of a stable process. *Z. Wahrscheinlichkeitstheorie verw. Gebiete 6*, (1966), 170-180.

31. H.F. TROTTER. A property of Brownian motion paths. *Ill. J. Math. 2*, (1958), 425-433.

M.T. BARLOW
Statistical Laboratory
16 Mill Lane,
Cambridge
CB2 1SB
U.K.

S.J. TAYLOR
Dept. of Mathematics
University of Virginia,
Charlottesville,
VA 22903
U.S.A.

E.A. PERKINS
Dept. of Mathematics
University of
British Columbia,
Vancouver,
B.C.
Canada V6T 1Y4

*Seminar on Stochastic Processes, 1984*
Birkhäuser, Boston, 1986

# NOTES ON THE INHOMOGENEOUS SCHRÖDINGER EQUATION

by

K. L. CHUNG*

In [1] and [2] we discussed the solution of the homogeneous Schrödinger equation $\left(\frac{\Delta}{2}+q\right)u = 0$ with boundary condition. It is customary in classical analysis to treat this problem as equivalent to the solution of the corresponding inhomogeneous equation $\left(\frac{\Delta}{2}+q\right)u = \phi$ with vanishing boundary condition, by a simple substitution. However, sufficient smoothness of the given data is required for this method. It turns out that the probabilistic approach is easily adapted to the inhomogeneous case, via the potentials. Relatively mild assumptions are sufficient for the purpose. Whereas it is possible to treat the problem in a "purely analytic" setting based on old and new Green's functions, we follow a different route and carry out the calculations by integrations over time rather than over space.

Let $D$ be a domain in $R^d$, $d \geq 1$, with $m(D) < \infty$, where $m$ is the Lebesgue measure in $R^d$. No regularity assumption is imposed on $\partial D$. Define a class of functions, to be denoted by $L^*(D)$, as follows: $\phi \in L^*(D)$ iff $\phi$ is locally bounded in $D$ and $\phi \in L^1(D,m)$. Then $L^*(D)$ is a linear space which admits the operation $\phi \to |\phi|$, and multiplication by a bounded measurable function.

Let $q$ be a bounded Borel measurable function on $R^d$, $Q = \sup_{x \in R^d} |q(x)|$;

*Research supported in part by NSF grant MCS83-01072 at Stanford University.

$\{X_t, t \geq 0\}$ the standard Brownian motion in $R^d$;

$$e_q(t) = \exp\left[\int_0^t q(X_s)ds\right];$$

and $\tau = \tau_D$ the first exit time from D. Define a semigroup $\{L_t^{(q)}, t \geq 0\}$ as follows: for positive Borel measurable f,

$$L_t^{(q)} f(x) = E^x\{t < \tau_D;\ e_q(t) f(X_t)\}.$$

The associated potential will be denoted by $V^{(q)}$:

$$V^{(q)} f = \int_0^\infty L_t^{(q)} f\, dt.$$

These notations are the same as in [2], except for the explicit indication of q. For $q \equiv 0$, $\{L_t^{(0)}\}$ reduces to the semigroup of the Brownian motion killed outside D; and $V^{(0)}$ becomes the classical Green's potential for D.

The gauge for (D,q) is defined in [1] to be the function $E^x\{e_q(\tau_D)\}$ for $x \in D$. It is proved that (Theorem 3.1 of [1]) the gauge is bounded in $\bar{D}$ if and only if $V^{(q)} 1 < \infty$ in D. We shall assume this condition throughout this note.

PROPOSITION 1. $V^{(q)}$ *maps* $L^*(D)$ *into* $L^*(D)$.

PROOF: If $\phi \in L^1(D,m)$, then by (4) and (7) of [2],

(1) $$\int_1^\infty L_t^{(q)} \phi\, dt \quad \text{is bounded in } D.$$

If $\phi \in L^*(D)$, then we have

(2) $$\int_0^1 L_t^{(q)}\phi\, dt \le e^Q \int_0^1 P_t(1_D\phi)dt.$$

For $d \ge 3$, the integral in the right member above is bounded by

(3) $$\int_D \frac{|\phi(y)|}{|x-y|^{d-2}}\, m(dy).$$

Let $B(x,r)$ denote the open ball with center $x$ and radius $r$. If $\delta > 0$, $B(x_0, 3\delta) \subset D$ and $x \in B(x_0,\delta)$, then the integral in (3) is bounded by

$$\int_{B(x,\delta)} \frac{M}{|x-y|^{d-2}}\, m(dy) + \frac{1}{\delta^{d-2}} \int_D |\phi(y)|m(dy)$$

where $M$ is a bound for $\phi$ in $B(x_0, 2\delta)$. This shows that the function of $x$ in (3) is locally bounded. It is integrable over $D$ because $\phi$ is, and

$$\sup_{y\in D} \int_D \frac{m(dx)}{|x-y|^{d-2}} < \infty.$$

Thus the function $x \to \int_0^1 L_t^{(q)}\phi(x)dt$ belongs to $L^*(D)$. Together with (1) we obtain

$$V^{(q)}\phi \in L^*(D).$$

For $d = 2$, the argument is similar if we replace $|x-y|^{2-d}$ by $\log^+|x-y|^{-1}$; for $d = 1$ the result is trivial. □

PROPOSITION 2. *For any* $\phi \in L^*(D)$, *we have*

(4) $$V^{(q)}\phi - V^{(0)}\phi = V^{(0)}(q\cdot V^{(q)}\phi) = V^{(q)}(q\cdot V^{(0)}\phi).$$

PROOF: Put $f \in V^{(q)}\phi$, then $f \in L^*(D)$ by Proposition 1; and also

$|q|V^{(q)}|\phi| \in L^*(D)$ since $q$ is bounded. Thus $V^{(0)}(|q|V^{(q)}|\phi|) \in L^*(D)$ by Proposition 1 with $q \equiv 0$, since $V^{(0)}1 < \infty$. We need the finiteness of $V^{(0)}(|q|V^{(q)}|\phi|)$ in the ensuing calculations to justify the change of order of integrations. By definition, we have

$$f(x) = E^x\{\int_0^\tau e_q(s)\phi(X_s)ds\}.$$

Substituting this into the second member below, we obtain the third by Markov property:

$$\begin{aligned}
V^{(0)}(qV^{(q)}\phi) &= E^x\{\int_0^\tau q(X_t)f(X_t)dt\} \\
&= E^x\{\int_0^\tau q(X_t)\int_0^{\tau\circ\theta_t} \exp[\int_0^s q(X_{t+r})dr]\phi(X_{t+s})dsdt\} \\
&= E^x\{\int_0^\tau q(X_t)\int_t^\tau e_q(s)e_q(t)^{-1}\phi(X_s)dsdt\} \\
&= E^x\{\int_0^\tau e_q(s)\phi(X_s)\int_0^s e_{-q}(t)q(X_t)dtds\} \\
&= E^x\{\int_0^\tau e_q(s)\phi(X_s)[1 - e_{-q}(s)]ds\} \\
&= E^x\{\int_0^\tau e_q(s)\phi(X_s)ds\} - E^x\{\int_0^\tau \phi(X_s)ds\} \\
&= V^{(q)}\phi(x) - V^{(0)}\phi(x).
\end{aligned}$$

This establishes the first equation in (4). To establish the second equation, we proceed as follows:

$$\begin{aligned}
V^{(q)}(q\cdot V^{(0)}\phi) &= E^x\{\int_0^\tau e_q(t)q(X_t)V^{(0)}\phi(X_t)dt\} \\
&= E^x\{\int_0^\tau e_q(t)q(X_t)\int_t^\tau \phi(X_s)dsdt\} \\
&= E^x\{\int_0^\tau \phi(X_s)\int_0^s e_q(t)q(X_t)dtds\}
\end{aligned}$$

$$= E^x\{\int_0^\tau \phi(X_s)[e_q(s) - 1]ds\}$$

$$= V^{(q)}\phi(x) - V^{(0)}\phi(x). \qquad \square$$

In order to pass from the potential operators to their infinitesimal generators, and obtain differential equations in the strict sense, we must impose smoothness conditions on $q$ as well as $\phi$. As in the Laplacian case (where $q \equiv 0$), we assume that $\phi$ is Hölder continuous in D, which implies its local boundedness; thus the additional hypothesis for $L^*(D)$ is now equivalent to the integrability of $\phi$ over D. We assume also that q is Hölder continuous as well as bounded in D. Recall $V^{(q)}1 < \infty$ by hypothesis.

PROPOSITION 3. *Under the assumptions stated above, we have*

$$\left(\frac{\Delta}{2} + q\right)V^{(q)}\phi = -\phi. \tag{5}$$

PROOF: We need the classical results that, if $\phi \in L^1(D,m)$, then $V^{(0)}\phi \in C^1(D)$; if $\phi$ is also Hölder continuous in D, then $V^{(0)}\phi \in C^2(D)$. Now since $\phi \in L^*(D)$, $V^{(q)}\phi \in L^*(D)$ by Proposition 1; hence, $qV^{(q)}\phi + \phi \in L^*(D)$; and consequently by (4),

$$V^{(q)}\phi = V^{(0)}(qV^{(q)}\phi + \phi) \in C^1(D). \tag{6}$$

Hence, $qV^{(q)}\phi + \phi$ is Hölder continuous in D, and therefore, by the same token, $V^{(q)}\phi \in C^2(D)$. Taking the Laplacian in (6), we obtain

$$\frac{\Delta}{2} V^{(q)}\phi = -qV^{(q)}\phi - \phi$$

which is (5). $\square$

When $\phi = q$ in the above, we obtain

$$(\frac{\Delta}{2} + q)(V^{(q)}q + 1) = 0.$$

Since

$$V^{(q)}q(x) + 1 = E^x\{e_q(\tau)\},$$

we retrieve the result that the gauge function is a solution of the homogeneous Schrödinger equation; see [1].

A class $K_d$ of unbounded $q$ has been considered which is said to be significant for mathematical physics. This class may be defined in our notation as follows: $q \in K_d$ iff

(7) $$\lim_{t \to 0} E^x\{\int_0^t 1_D|q|(X_s)ds\} = 0$$

uniformly in $x \in \bar{D}$. It follows that, if $d \geq 3$, then

(8) $$\int_D \frac{|q|(y)}{|x-y|^{d-2}} m(dy) \quad \text{is bounded in } x \in \bar{D}.$$

Furthermore, we know that the finiteness of the gauge for $(D,q)$ implies that

(9) $$V^{(q)}1 \quad \text{is bounded in } D,$$

(10) $$V^{(q)}|q| \quad \text{is bounded in } D.$$

We should also point out that even for an unbounded function in the class $K_d$, the "obvious" integration formula

$$\int_0^t e_q(s)q(X_s)ds = e_q(t) - 1$$

remains operable. This is because under (7), $q(X_s)$ is finite for

(m) a.e. s, and we can check by the classic criterion that $e_q(t)$ is an absolutely continuous function of $t$. It follows from (9) that $V^{(q)}$ maps $L^\infty(D)$ into itself. Next, if $\phi \in L^\infty(D)$, then by (9)

$$V^{(0)}(|q|V^{(q)}|\phi|) \le \| V^{(0)}|q| \| \cdot \| V^{(q)}1 \| \cdot \| \phi \|,$$

and a similar inequality when $V^{(0)}$ and $V^{(q)}$ are interchanged above, using (10). Thus the calculations in the proof of Proposition 2 remain valid.

The extension of these results to $\phi \in L^*(D)$ seems more difficult. I am indebted to Zhongxin Zhao for the discussion below. We begin with an inequality due to B. Simon [3, Theorem 13.7.2 (2)]. Let $v^{(q)}$ denote a density for the kernel $V^{(q)}$; its existence follows from general principles. If $d \ge 3$, $q \in K_d$, and the gauge for $(D,q)$ is finite, then

$$v^{(q)}(x,y) \le \frac{C}{|x-y|^{d-2}} . \tag{11}$$

Strictly speaking, this result was proved by Simon under the alternative assumption that the maximum eigenvalue for the operator $\frac{\Delta}{2} + q$ is strictly negative. This assumption can be shown by the methods of [1] and [2] to be equivalent to the finiteness of the gauge for $(D,q)$. Note that (11) is trivial if $q \equiv 0$ and easy if $q$ is bounded. Recall that the function of $x$ given in (3) has been shown to belong to $L^*(D)$. It follows from (11) that $V^{(q)}\phi \in L^*(D)$. Thus Proposition 1 is true for $q \in K_d$.

Next we have by (11), for $\phi \in L^*(D)$, $\phi \ge 0$:

$$V^{(q)}(|q|V^{(0)}\phi)(x) \le C^2 \int_D \int_D \frac{|q|(y)\phi(z)}{|x-y|^{d-2}|y-z|^{d-2}} m(dy)m(dz). \tag{12}$$

Using the elementary inequality

$$\frac{1}{|x-y|^{d-2}|y-z|^{d-2}} \le \left(\frac{1}{|x-y|^{d-2}} + \frac{1}{|y-z|^{d-2}}\right)\frac{2^{d-2}}{|x-z|^{d-2}}$$

and the property (8), we see that the double integral in (11) is less than a constant multiple of $\int_D \phi(z)|x-z|^{2-d}\, m(dz)$ which belongs to $L^*(D)$ as shown in the proof of Proposition 1 above. Thus the left member of (11) is finite for all $x \in D$. Similarly $V^{(0)}(|q|V^{(q)}\phi)(x) < \infty$ for all $x \in D$. These conclusions justify the application of Fubini's theorem in the calculations made in the proof of Proposition 2, and the results follow exactly as before.

References

1. K. L. CHUNG and K. M. RAO. Feynman-Kac functional and the Schrödinger equation. *Seminar on Stochastic Processes, 1981*, 1-29. Birkhäuser, Boston, 1981.

2. K. L. CHUNG. An inequality for boundary value problems. *Seminar on Stochastic Processes, 1982*, 111-122. Birkhäuser, Boston, 1983.

3. BARRY SIMON. Schrödinger semigroups. *Bull. Amer. Math. Soc.* 7 (1982), 447-526.

K. L. CHUNG
Department of Mathematics
Stanford University
Stanford, California 94305

*Seminar on Stochastic Processes, 1984*
Birkhäuser, Boston, 1986

# GAUGE THEOREM FOR THE NEUMANN PROBLEM

by

K.L. CHUNG* and PEI HSU*

Let D be a bounded domain in $\mathbb{R}^d$ and let $(\Delta/2 + q)u = 0$ be Schrödinger's equation on D. The Dirichlet problem for the equation was studied first in [2] for bounded q and then in [1] and [4] for $q \in K_d$ (see below for definition). The gauge function for the Dirichlet problem is defined in [2] as

$$(1) \qquad G(x) = E^x[\exp(\int_0^{\tau_D} q(B_s)\,ds)],$$

where $B = \{B_t, t \geq 0\}$ is the standard Brownian motion on $\mathbb{R}^d$ and $\tau_D$ is the first exit time of D. One striking property of the gauge function proved in [2] and [4] is the following

THEOREM 1. *If* G *is not identically infinite, then it is bounded on* $\bar{D}$.

The gauge function plays a key role in the solution of the Dirichlet problem; see the references mentioned above.

In this paper, we define a gauge function which plays a role in the Neumann problem similar to that of the gauge in the Dirichlet

* Research supported in part by NSF grant MCS83-01072 at Stanford University.

problem. It turns out that this new function also has the property stated in Theorem 1.

To define this gauge function, let us start with the definition of the class $K_d$.

DEFINITION. A measurable function q is in the class $K_d$ iff

(2) $$\lim_{\alpha\to 0} \sup_{x\in R^d} \int_{|x-y|\le\alpha} G_d(x,y)\,|q|(y)\,dy = 0,$$

where $G_d$ is the fundamental solution of Laplace's equation in $R^d$, namely

$$G_d(x,y) = \begin{cases} |x-y|, & \text{if } d = 1; \\ \log|x-y|^{-1}, & \text{if } d = 2; \\ |x-y|^{-2+d}, & \text{if } d \ge 3. \end{cases}$$

It is proved in [1] that condition (2) is equivalent to the condition

(3) $$\lim_{t\to 0} \sup_{x\in R^d} \int_0^t \int_{R^d} |q|(y)\Gamma(s,x,y)\,dy\,ds = 0,$$

where $\Gamma(t,x,y)$ is the transition density function of the standard Brownian motion on $R^d$:

$$\Gamma(t,x,y) = (2\pi t)^{-d/2} e^{-\|x-y\|^2/2t}.$$

Let D be a bounded domain with $C^3$ boundary. Let $X = \{X_t,\ t \ge 0\}$ be the standard reflecting Brownian motion on D and $L = \{L(t),\ t \ge 0\}$ be its boundary local time. We refer to [3] for a discussion of reflecting Brownian motion and the boundary local time.

The condition that the restriction of q on D lies in the class

$K_d$ can be characterized in terms of reflecting Browniàn motion.

THEOREM 2. *Let* q *be measurable on* $\mathbb{R}^d$, *then* $qI_D \in K_d$ *if and only if*

$$\lim_{t \to 0} \sup_{x \in \bar{D}} E^x \left[ \int_0^t |q|(X_s)\, ds \right] = 0 \tag{4}$$

We delay the proof of Theorem 2 to the end of the paper.

Let $qI_D \in K_d$ and put

$$e_q(t) = \exp\left[ \int_0^t q(X_s)\, ds \right].$$

This is finite a.s. by (4). Now we define

$$G_q(x) = E^x\left[ \int_0^\infty e_q(s)\, dL(s) \right]. \tag{5}$$

$G_q$ will be called the gauge function for the Neumann problem.

Define the semigroup $\{R_t^{(q)},\ t \geq 0\}$ as follows:

$$R_t^{(q)} f(x) = E^x[e_q(t) f(X_t)].$$

Observe that this semigroup is not necessarily sub-Markovian. In the following, $A_t$ and $C_t$ denote constants depending on t. They are not necessarily the same at each appearance.

LEMMA 1. *For any fixed* $t > 0$, *there is a constant* $A_t$ *such that*

$$\forall f \in L^1(D):\quad \|R_t^{(q)} f\|_\infty \leq A_t \|f\|_1. \tag{6}$$

PROOF. The proof is the same as that for killed Brownian motion given in [1]. The conditions used there are also satisfied by the reflecting Brownian motion. □

Lemma 1 is used in the next lemma to obtain an inequality in the opposite direction.

LEMMA 2. *Let* $f \geq 0$ *be measurable on* $\bar{D}$. *For any fixed* $t > 0$, *there is a constant* $A_t$ *such that*

$$\forall x \in \bar{D}: \quad \int_{\bar{D}} f(y)dy \leq A_t R_t^{(q)} f(x). \tag{7}$$

PROOF. By (6), with $-q$ for $q$, and the Schwarz inequality,

$$\begin{aligned} E^x[f(X_t)]^2 &= E^x[\, e_{\frac{1}{2}q}(t)\; f^{\frac{1}{2}}(X_t)\; e_{-\frac{1}{2}q}(t)\; f^{\frac{1}{2}}(X_t)]^2 \\ &\leq E^x[e_q(t)\; f(X_t)]\, E^x[e_{-q}(t)\; f(X_t)] \\ &= R_t^{(q)} f(x)\; R_t^{(-q)} f(x) \\ &\leq A_t\; \|f\|_1\; R_t^{(q)} f(x). \end{aligned} \tag{8}$$

On the other hand, for any $t > 0$, there is a positive constant $C_t$ such that

$$\forall (x,y) \in \bar{D} \times \bar{D}: \quad p(t,x,y) \geq C_t,$$

where $p(t,x,y)$ is the transition density function of the reflecting Brownian motion $X$. Thus,

$$E^x[f(X_t)] \geq C_t \int_{\bar{D}} f(y)dy. \tag{9}$$

By (8) and (9),

$$C_t^2\; \|f\|_1^2 \leq A_t\; \|f\|_1\; R_t^{(q)} f(x). \tag{10}$$

Hence, if $\|f\|_1 < \infty$, then

$$\|f\|_1 \le A_t \, R_t^{(q)} f(x). \tag{11}$$

In general, we can replace f by $f \wedge n$ in (11) and apply the monotone convergence theorem. The lemma is proved. □

THEOREM 3. *Let* $q \in K_d$ *and* $G_q$ *be the gauge function defined in* (5). *If* $G_q \not\equiv \infty$, *then it is continuous on* $\bar{D}$, *hence bounded on* $\bar{D}$.

PROOF. By the Markov property,

$$\infty \ge G_q(x) = E^x\Big[\int_0^t e_q(s)\, dL(s)\Big] + E^x[e_q(t) G_q(X_t)]. \tag{12}$$

For any fixed $t > 0$,

$$E^x\Big[\int_0^t e_q(s)\, dL(s)\Big]^2 \le E^x[e_{|q|}(t)\, L(t)]^2 \le E^x[e_{2|q|}(t)]\; E^x[L(t)^2]. \tag{13}$$

The first factor in the last member of (13) is bounded by Khas'minskii's lemma (see [1]) and Theorem 2. It is easy to show that

$$\sup_{x\in\bar{D}} E^x[L(t)^2] \le 2\big(\sup_{x\in\bar{D}} E^x[L(t)]\big)^2. \tag{14}$$

Since

$$E^x[L(t)] = \int_0^t \int_{\partial D} p(s,x,y)\, d\sigma(y)\, ds \le \int_0^t \frac{C}{\sqrt{t}}\, ds = 2C\sqrt{t}$$

for a constant C (see [3]), it follows that the second factor in the last member of (13) is also bounded. Hence

$$E^x\Big[\int_0^t e_q(s)\, dL(s)\Big] \text{ is bounded on } \bar{D}. \tag{15}$$

Now suppose $G_q(x_0) < \infty$. By (12),

$$\infty > G_q(x_0) \geq R_t^{(q)} G_q(x_0).$$

By Lemma 2,

$$R_t^{(q)} G_q(x_0) \geq A_t \, \|G_q\|_1.$$

Hence $\|G_q\|_1 < \infty$. By Lemma 1,

$$\|R_t^{(q)} G_q\|_\infty \leq A_t \, \|G_q\|_1.$$

This shows that $R_t^{(q)} G_q$ is bounded. It follows from (12) and (13) that $G_q$ is bounded on $\bar{D}$.

Furthermore, it is known that the semigroup $\{R_t^{(q)}, t > 0\}$ is strong Feller, hence $R_t^{(q)} G_q$ is continuous. Now it follows from (13) and (14) that

$$\lim_{t \to 0} \sup_{x \in \bar{D}} E^x\left[\int_0^t e_q(s) \, dL(s)\right] = 0. \tag{16}$$

Hence by (12), $G_q$ is continuous on $\bar{D}$. The theorem is proved. □

It remains to complete the

PROOF OF THEOREM 2. It was proved in [3] that the transition density function $p(t,x,y)$ of the standard reflecting Brownian motion on D can be written in the form

$$p(t,x,y) = p_0(t,x,y) + p_1(t,x,y), \tag{17}$$

where $p_0$ and $p_1$ have the following properties.

(a) There are positive constants $c_1$, $c_2$, and a such that

(18) $$c_2\ \Gamma(t,x,y) \leq p_0(t,x,y) \leq c_1\ \Gamma(at,x,y).$$

(b) $p_1(t,x,y)$ has the form

$$p_1(t,x,y) = \int_0^t \int_D p_0(t-u,x,z)\ f(u,z,y)\ dz\ du,$$

with

$$\sup_{y\in R^d} \int_D |f(t,x,y)|\ dx \leq \frac{C}{\sqrt{t}},$$

where C is a constant.

Now let

$$M_0(t) = \sup_{z\in R^d} \int_0^t \int_D p_0(s,y,z)\ |q|(y)\ dy\ ds.$$

We have

(19) $$\int_0^t \int_D |p_1(s,y,x)|\ |q|(y)\ dy\ ds$$

$$\leq \int_0^t \int_D \int_0^s \int_D p_0(s-u,y,z)\ |f(u,z,x)|\ |q|(y)\ dz\ du\ dy\ ds$$

$$\leq M_0(t) \sup_{x\in\bar{D}} \int_0^t \int_D |f(u,z,x)|\ dz\ ds$$

$$\leq 2C\sqrt{t}\ M_0(t).$$

By the symmetry of $p(t,x,y)$ in $(x,y)$, we have

$$E^x[\int_0^t |q|(X_s)\ ds] = \int_0^t \int_D p(s,y,x)\ |q|(y)\ dy\ ds$$

$$= \int_0^t \int_D p_0(s,y,x)\ |q|(y)\ dy\ ds$$

$$+ \int_0^t \int_D p_1(s,y,x)\ |q|(y)\ dy\ ds.$$

The absolute value of the last term is not greater than $M_0(t)/2$ if

$t \leq 1/8C^2$. Hence for $t \leq 1/8C^2$, we have

$$\frac{1}{2}M_0(t) \leq \sup_{x\in\bar{D}} E^x[\int_0^t |q|(X_s)\, ds] \leq \frac{3}{2}M_0(t). \tag{20}$$

On the other hand, as recalled before, $qI_D \in K_d$ if and only if (3) holds. By (18), the latter condition is equivalent to the condition $\lim_{t\to 0} M_0(t) = 0$. Hence by (20), $qI_D \in K_d$ is equivalent to (4). The theorem is proved.

We refer to [3] for further properties of the gauge function $G_q$ as well as its application to the Neumann problem.

## References

1. M. AIZENMAN and B. SIMON. Brownian Motion and Harnack's inequality for Schrödinger Operators. *Comm. Pure Appl. Math.* *35*(1982), 209-271.

2. K.L. CHUNG and K.M. RAO. Feynman-Kac Functional and the Schrödinger Equation. *Seminar on Stochastic Process 1981*, pp. 1-29. Birkhäuser, Boston, 1981.

3. PEI HSU. Reflecting Brownian Motion, Boundary Local Time and the Neumann Problem. Doctoral Dissertation, Stanford University, June 1984.

4. Z.X. ZHAO. Conditional Gauge with Unbounded Potentials. Z. *Wahrscheinlichkeitstheorie verw. Geb.* *35*(1983), 13-18.

K.L. CHUNG
Department of Mathematics
Stanford University
Stanford, CA 94305

Pei HSU
Courant Institute of Mathematics
New York University
251 Mercer Street
New York, NY 10012

*Seminar on Stochastic Processes, 1984*
Birkhäuser, Boston, 1986

# QUASI-STATIONARY DISTRIBUTIONS, EIGENMEASURES, AND EIGENFUNCTIONS OF MARKOV PROCESSES*

by

JOSEPH GLOVER

## 0. Introduction

Let $P_t$ be a submarkov semigroup on a Lusin topological state space E with Borel field $\mathcal{E}$. We call a positive sigma-finite measure m on E an *eigenmeasure* if $mP_t = e^{-ct}m$ for some real number c. We call a positive $\mathcal{E}$-measurable function f an *eigenfunction* if $P_tf = e^{-ct}f$ for some real number c. In each case, we call c the eigenvalue of either the eigenmeasure or the eigenfunction. Eigenmeasures are also known by the name quasi-stationary distributions in the Markov chain literature: see [5], [17], [18].

We have two main aims in this paper. First, we try to find simple conditions guaranteeing existence of eigenmeasures and eigenfunctions. Early work on eigenmeasures was motivated through applications to genetics of Markov chains on discrete state spaces (see [17], [18]), and the problem of existence essentially reduces to a problem about transition matrices. In this approach, the eigenmeasures are often obtained as various conditional limiting distributions involving

*Research supported in part by NSF Grant DMS-8318204.

the semigroup of the original process ([5],[18]). For example, if $X(t)$ is a Markov chain on a finite state space $E_\Delta = \{1,2,3,\ldots,n,\Delta\}$, let $Q$ be the (sub)markov transition matrix of $X(t)$ on $E$. Assume $Q$ is indecomposable and non-cyclic. Then there is a unique largest positive eigenvalue $p$ with a positive left eigenvector $v$ and a positive right eigenvector $w$. Normalize $v$ and $w$ so that

$$\sum_{i=1}^{n} v_i = 1 \quad \text{and} \quad \sum_{i=1}^{n} w_i v_i = 1 .$$

Then $v_j$ and $w_j v_j$ may be obtained as

$$(0.1) \qquad v_j = \lim_{n\to\infty} P^i[X(n)=j \mid n < \zeta]$$

$$w_j v_j = \lim_{m\to\infty} \lim_{n\to\infty} P^i[X(m)=j;\ m < n \mid \zeta < n]$$

for any $i$ in $E$. Here, $\zeta \doteq \inf\{n: X(n)=\Delta\}$. We refer to representations such as those given in (0.1) as "concrete". Note that they can be written in terms of the semigroup of the process $X(t)$. We are able to find fairly simple conditions guaranteeing existence of eigenmeasures and eigenfunctions by applying fixed point theorems of a more (2.7) or less (1.10) constructive nature. Once we know that these exist, it is often possible to give a "concrete" method of obtaining it or another one. The problem has been investigated in detail on finite and countable state spaces (see the bibliographies in the references listed above), but less work has been done for the case of continuous state spaces. Motivated by D. Sullivan's work [20] with the heat equation on manifolds, D. Stroock [19] has examined the problem of existence of eigenmeasures from a potential theory viewpoint. Sections 1 through 4 are intended to complement his work.

The second aim of this paper is to reformulate recent work of Chung and Rao [3] in an eigenfunction setting. They considered the problem of minimizing the energy of a nonsymmetric potential kernel on a compact set K and showed that a modification of the equilibrium measure of K minimizes a modification of the potential kernel. We examine the nature of these modifications in section 5 and show in section 6 that their results are in fact special cases of interesting results about eigenfunctions. Their arguments are modified in only simple ways in section 6 to produce the results; one need only realize that their arguments can generalize to eigenfunctions, and the rest is easy.

Why should probabilists be interested in eigenmeasures and eigenfunctions? Sufficient reasons are supplied in the references mentioned above and in the body of the paper, but let us mention one more which does not appear explicitly later in this paper. For the remainder of the introduction, we assume that $P_t$ is the semigroup of a right Markov process X(t). Let m be an eigenmeasure with eigenvalue $c > 0$. If we let

$$\mathcal{G} = \sigma\{f(X_\zeta)_- : f \text{ is } \alpha\text{-excessive}\},$$

then $\mathcal{G}$ and $\zeta$ are $P^m$-independent. To see this, let g be any finite product of $\alpha$-excessive functions and consider

$$P^m[g(X_\zeta)_- : t < \zeta] = P^m[(g(X_\zeta)_-)\circ\theta_t ; t < \zeta] =$$

$$P^m[P^{X(t)}[g(X_\zeta)_-]] = e^{-ct}P^m[g(X_\zeta)_-] = P^m[t < \zeta]P^m[g(X_\zeta)_-]$$

The assertion now follows by applying the monotone class theorem. For example, if X(t) is Brownian motion killed when it leaves a bounded open set E, let f be the first eigenfunction of $-\frac{1}{2}\Delta$ on E (with Dirichlet boundary conditions). Then $dm = f(x)dx$ is an eigenmeasure with eigenvalue c, where $cf = -\frac{1}{2}\Delta f$. In fact, the analogue of (0.1) for Brownian motion can be derived analytically from Mercer's theorem (see [13] and [16]). More generally, analogous formulae can be obtained for $L^2$-symetric semigroups: see [19]. Therefore, most of our attention is devoted to the nonsymmetric case. In general, eigenmeasures and eigenfunctions need not exist. When they do, it may happen that the eigenvalues are negative. For example, if $X(t)$ is uniform motion to the right on $(0,\infty)$, then $X(t)$ has no nonzero eigenmeasures since $P_t$ is concentrated on $[t,\infty]$. If $X(t)$ is uniform motion to the right on $(-\infty,\infty)$, then each real c is an eigenvalue of an eigenfunction $e^{-cx}$. In particular, eigenfunctions need not be excessive functions: they are excessive if and only if $c > 0$. These examples are trivial, but after all, their behavior is reflected in that of space-time processes, so they are worth mentioning.

In section 1, we consider the case of a Ray resolvent on a compact metric space $E_\Delta$ with $\Delta$ isolated in E. We use Stroock's approach to defining the "first" eigenvalue (1.6) and show that it is the eigenvalue of a finite eigenmeasure. The Schauder fixed point theorem is used in the proof. The last theorem (1.16) in section 1 gives a sufficient condition for existence of an eigenfunction corresponding to the first eigenvalue of the dual semigroup. It relies on the following observation ([19]). Let $X(t)$ and $\hat{X}(t)$ be two right processes in duality with respect to an excessive measure m as described in Chapter VI of [2]. Then $\lambda$ is an eigenmeasure for $\hat{X}$ if and only if $f = d\lambda/dm$ is an eigenfunction for X. Note that f may be chosen to be excessive.

In section 2, we consider semigroups $P_t(x,\cdot) \ll m$, and we give a necessary and sufficient condition for $P_t$ to have a bounded eigenfunction. This result seems to be most useful in showing that if a semigroup has a bounded eigenfunction, then mild perturbations of it obtained by killing still have bounded eigenfunctions (see Ex. (2.8)). The fixed point theorem used in this section gives a concrete method of obtaining the eigenfunction when it is unique (up to constant multiples).

In section 3, we give one simple situation in which uniqueness holds. Let E be compact, and let $(U^\alpha)$ be a strong Feller resolvent so that whenever $B \in \mathcal{E}$, either $U^\alpha 1_B=0$ on E or $U^\alpha 1_B>0$ on E. Then there is at most one eigenfunction with eigenvalue p. In section 4, we note that Yosida's ergodic theorem may be applied in certain situations to yield a concrete procedure for obtaining an eigenfunction.

In section 5, we recall the Chung-Rao [3] energy results and show that the measure $\pi$ they construct is an invariant measure of a process obtained from X in a natural way. In section 6, we consider the case of two processes X and $\hat{X}$ in duality with respect to m with eigenfunctions f and $\hat{f}$, respectively. We show under certain hypotheses that $dm^\circ = f\hat{f}\, dm$ minimizes the energy of the modified potential kernel $u(x,y)/f(x)\hat{f}(y)$.

Notation is standard and can be found in [2] and [7]. In particular, if W is a metric space, C(W) (resp. $C(W)^+$) denotes the collection of bounded continuous functions on W (resp. and which are positive).

## 1. Eigenmeasures of Ray Resolvents

Let $E_\Delta$ be a compact metric space with Borel field $\mathcal{E}_\Delta$. We assume that $E_\Delta$ contains an *isolated* point $\Delta$, and we set $E = E_\Delta - \{\Delta\}$. Let $(U^\alpha)_{\alpha>0}$ be a Ray resolvent on $E_\Delta$. That is $U^\alpha : C(E_\Delta) \to C(E_\Delta)$, and the bounded $\alpha$-supermedian functions separate points on $E_\Delta$. In addition, we assume that

$$(1.1)\qquad \alpha U^{\alpha} 1_{\{\Delta\}}(\Delta) = 1 \quad \text{for every } \alpha > 0, \text{ and}$$

$$(1.2)\qquad U^{\alpha} 1_{E}(x) > 0 \quad \text{for every } x \text{ in } E.$$

Using only these hypotheses, we show that $(U^{\alpha})$ has a finite positive eigenmeasure. The Ray-Knight compactification procedure can be applied to any right process as in [7] to yield a Ray process on a compactification $E_{\Delta}$ of its original state space $F_{\Delta}$. By adding $1_F$ to the Ray cone, one can ensure that $\Delta$ is isolated, and (1.1) will be satisfied. However, (1.2) will often not be satisfied.

We make the following definitions.

$$(1.3)\qquad M = \{\text{finite positive measures } \mu \text{ on } E\}$$

$$(1.4)\qquad C(p) = \{\mu \in M\colon \alpha\mu U^{\alpha-p} 1_{E} \leq \mu \text{ for every } \alpha > p\}$$

$$(1.5)\qquad K(p) = \{\mu \in C(p)\colon \mu(E) = 1\}$$

$$(1.6)\qquad q = \sup\{p \in R\colon K(p) \neq \emptyset\}$$

These definitions were given by Stroock [19], as was the next proposition (although with slightly different hypotheses). There, he shows that (1.6) is the correct definition of the "first" eigenvalue. Before continuing, recall that every Ray resolvent is generated by a right continuous Ray semigroup $P_t$, and (1.4) is equivalent to

$$(1.7)\qquad \{\mu \in M\colon \mu P_t 1_E \leq e^{-pt}\mu \quad \text{for every } t > 0\}$$

It is more natural to use (1.6) in this section.

(1.8) PROPOSITION. (i) $K(p)$ *is compact for every* $p$ *(in the vague topology).*

(ii) $q \geq 0$ *and* $K(q) \neq \emptyset$.

(iii) $q < \infty$.

PROOF. (i) Fix $p$, and choose a sequence $(\mu(n)) \subset K(p)$. Since $\Delta$ is isolated in $E_\Delta$, $E$ is compact, so $(\mu(n))$ contains a subsequence $(\mu(n_k))$ converging vaguely to a probability $\mu$ on $E$. Since $\mu(n_k)$ is in $K(p)$,

$$\alpha\mu(n_k)U^{\alpha-p}(f1_E) \leq \mu(n_k)(f)$$

for every $f$ in $C(E)^+$. Since $U^{\alpha-p}(f1_E)$ is also in $C(E)$ for every $\alpha > p$, we may pass to the limit to obtain

$$\alpha\mu U^{\alpha-p}(f1_E) \leq \mu(f) .$$

Therefore, $\mu \in K(p)$.

(ii) If $p < 0$, choose $x$ in $E$, and set $\mu = \varepsilon_x U^{-p}1_E$. By (1.2), $\mu \neq 0$ on $E$. By the resolvent equation, if $\alpha > p$,

$$\alpha\mu U^{\alpha-p}1_E = \alpha\varepsilon_x U^{-p}1_E U^{\alpha-p}1_E = \varepsilon_x U^{-p}1_E - \varepsilon_x U^{\alpha}1_E \leq \mu.$$

Therefore, $q \geq 0$. Let $p(n)$ increase to $q$. For each $n$, $K(p(n))$ is nonempty, and we choose a measure $\mu(n)$ therein. Since $E$ is compact, $(\mu(n))$ contains a subsequence $\mu(n_k)$ converging vaguely to a probability $\mu$ on $E$. If $f \in C(E)^+$ and $\alpha > q$,

$$\alpha\mu(n_k)U^{\alpha-p(n_k)}f \leq \mu(n_k)(f) .$$

Now

$$(1.9)\quad |\mu(n_k)U^{\alpha-p(n_k)}f - \mu U^{\alpha-q}f|$$

$$\le \mu(n_k)[|U^{\alpha-p(n_k)}f - U^{\alpha-q}f|] + |\mu(n_k)U^{\alpha-q}f - \mu U^{\alpha-q}f|.$$

The first term on the right hand side of (1.9) is less than

$$\|f\|_\infty |(p(n_k) - \alpha)^{-1} - (q - \alpha)^{-1}| ,$$

which converges to zero. Since $U^{\alpha-q}f \in C(E)$, the second term goes to zero. Therefore, we obtain $\alpha\mu U^{\alpha-q}f \le \mu(f)$, so $\mu \in K(q)$.

(iii) If $\mu \in K(q)$, then $(1 + q)\mu U^1 1_E \le \mu(E) < \infty$. By (1.2), $\mu U^1 1_E > 0$, so $q < \infty$. □

Now recall the statement of the Schauder fixed point theorem.

(1.10) THEOREM [14]. *Let* C *be a nonempty compact convex subset of a locally convex space. If* T *is a continuous map from* C *into* C, *then* T *has a fixed point in* C.

This theorem is purely an existence theorem; it gives no hint, in general, about how to find a fixed point. Since we use it in the next theorem to prove existence of an eigenmeasure, (1.11) is also purely an existence result, and we shall need to examine more "concrete" methods for finding eigenmeasures later. A similar, but simpler, application of (1.10) may be found in [3].

(1.11) THEOREM. *Let* $(U^\alpha)$ *be a Ray resolvent on* E *satisfying* (1.1) *and* (1.2). *There is a probability* $\mu$ *in* K(q) *so that* $(\alpha+q)\mu U^\alpha 1_E = \mu$ *for every* $\alpha > 0$.

PROOF. Define a map $T:K(q) \to K(q)$ by setting

$$T(\nu)f = \frac{\nu U^1(f1_E)}{\nu U^1(1_E)} \tag{1.12}$$

for every f in C(E). If $\nu_n$ is a sequence of measures in K(q) converging vaguely to $\nu$ in K(q), then $\nu_n U^1(f1_E)$ converges to $\nu U^1(f1_E)$ since $U^1(f1_E)$ is continuous. Since $1_E$ is continuous on $E_\Delta$, $U^1 1_E$ is continuous on $E_\Delta$, so $U^1 1_E > c > 0$ on E by (1.2). Therefore, $\nu_n U^1 1_E$ converges to $\nu U^1 1_E > 0$. Thus $T(\nu_n)$ converges vaguely to $T(\nu)$, so T is continuous on K(q) in the vague topology. The collection of finite measures on E endowed with the vague topology is a locally convex space, and we have shown (1.8) that K(q) is nonempty and compact. It is easy to check that K(q) is also convex. By the Schauder fixed point theorem, there is a measure $\mu$ in K(q) with $T(\mu) = \mu$. That is, if we let $c = \mu(U^1 1_E)$, then $c\mu = \mu U^1 1_E$. The following simple argument from [2, chapter V,(5.10)] shows that

$$(\alpha - 1 + \tfrac{1}{c})\, \mu U^\alpha 1_E = \mu. \tag{1.13}$$

The resolvent equation gives for any $\alpha$ that $U^\alpha[I - (1-\alpha)U^1] = U^1$, so

$$U^\alpha = \sum_{n=0}^{\infty} (1-\alpha)^n (U^1)^{n+1} \tag{1.14}$$

provided $|1 - \alpha| < 1$. If $f \in C(E)$,

$$\mu U^\alpha f = \mu \sum_{n=0}^{\infty} (1-\alpha)^n (U^1)^{n+1} f$$

$$= \sum_{n=0}^{\infty} (1-\alpha)^n \mu (U^1)^{n+1} f$$

$$= \sum_{n=0}^{\infty} (1-\alpha)^n c^{n+1} \mu(f) = \frac{c}{1-c(1-\alpha)} \mu(f) ,$$

proving (1.13) if $|1-\alpha| < 1$. The case of general $\alpha$ now follows from the resolvent equation. To complete the proof, we need only show that $c^{-1} - 1 = q$. Since $\mu$ is in $K(q)$,

$$(\alpha + q) \mu U^{\alpha} 1_E \leq \mu = (\alpha + c^{-1} - 1) \mu U^{\alpha} 1_E ,$$

so, $c^{-1} - 1 \geq q$. Since $\mu$ is in $K(c^{-1} - 1)$, we have $K(c^{-1} - 1) \neq \emptyset$; so, $q \geq c^{-1} - 1$ by definition of $q$ (1.6). Therefore, $q = c^{-1} - 1$. □

Theorem (1.11) covers the case where E is a finite discrete state space. This situation has been explored in a number of articles [5], [17], [18].

Recall that each Ray resolvent is the resolvent of a Ray semigroup $P_t$, [7]. That is, $P_t$ is a Borel measurable semi-group of a right continuous Ray process on E. Thus we have the following result by inverting Laplace transforms.

(1.15) COROLLARY. *If* $(P_t)$ *is the Ray semigroup with resolvent* $(U^{\alpha})$, *then there is a probability* $\mu$ *in* $K(q)$ *so that* $\mu P_t 1_E = e^{-qt} \mu$.

At this point, it may be worth pointing out a result which is dual to the one given in (1.11). Once again, let $E_{\Delta}$ be compact, and let $\Delta$ be isolated in E . Here by strong Feller, we mean $\hat{U}^1 : b\mathcal{E} \to C(E)$.

(1.16) THEOREM. *Suppose* $(U^{\alpha})$ *and* $(\hat{U}^{\alpha})$ *are two resolvents of right processes on* $E_{\Delta}$ *which are in strong duality with respect to a sigma finite excessive reference measure* m *as described in Chapter* VI *of* [2].

*Assume* $(\hat{U}^\alpha)$ *satisfies* (1.1) *and* (1.2). *Let* $\hat{q}$ *be defined for* $(\hat{U}^\alpha)$ *as in* (1.6). *If* $\hat{U}^1$ *is strong Feller, then there is a positive function* f *on* E *so that* $(\alpha + \hat{q})U^\alpha f(x) = f(x)$ *for every* $\alpha > 0$ *and for every* x *in* E. *That is,* f *is an eigenfunction of* $(U^\alpha)$ *with eigenvalue* $\hat{q}$.

PROOF. By (1.11), there is a probability $\mu$ in $K(\hat{q})$ so that $(\alpha + \hat{q})\mu\hat{U}^\alpha 1_E = \mu$ for every $\alpha > 0$. But this shows $\mu \ll m$. If $f^\circ$ is a version of the Radon-Nikodym derivative $d\mu/dm$, then $f(x) = \lim_{\alpha\to\infty} \alpha U^\alpha f^\circ(x)$ is the desired eigenfunction. □

## 2. Existence of Bounded Eigenfunctions of Absolutely Continuous Semigroups

Let $E_\Delta$ be a U-space (i.e. $E_\Delta$ is homeomorphic to a universally measurable subset of a compact metric space) with Borel field $\mathcal{E}_\Delta$. Let $(P_t)$ be the semigroup of a right process on $E_\Delta$ so that

(2.1) $$P_t 1_{\{\Delta\}}(\Delta) = 1 \text{ for all } t > 0$$

(2.2) $$P_t(x,\cdot) \ll m \text{ for every } x \text{ and for every } t > 0.$$

We may assume that m is 1-excessive (for, if not, replace m with $\lambda U^1$, where $\lambda$ is a probability on $E_\Delta$ which is equivalent to m). Then we may choose a density $p_t(x,y)$ so that $P_t(x,dy) = p_t(x,y)m(dy)$. No assumptions concerning the existence of a dual process are made now. Let

$$L^1 = L^1(m) = \{h: \int |h|\, dm < \infty\} .$$

Let $L^\infty = L^\infty(m)$ be the dual of $L^1$ equipped with the weak*-topology: $g_n \to g$ in $L^\infty$ if and only if $\int g_n h\, dm \to \int gh\, dm$ for every h in $L^1$. We let $||g|| = \text{ess sup } |g|$. By Alaoglu's theorem, the unit ball $B_1$ in $L^\infty$

is compact in this topology, and if we set

$$B_1^+ = \{f \in B_1: \ f \geq 0 \ \text{m-a.e.}\} = \bigcap_{0 \leq h \in L^1} \{f \in B_1: \int fh \, dm \geq 0\},$$

then $B_1^+$ is compact. Define

$$(2.3) \qquad h\hat{P}_t(y) = \int h(x) \, p_t(x,y) \, m(dx)$$

$$(2.4) \qquad B(p) = \bigcap_{t>0} \bigcap_{0 \leq h \in L^1} \{f \in B_1^+: \ e^{pt} \int h\hat{P}_t \cdot f \, dm \leq \int hf \, dm\}$$

Since $m(h\hat{P}_t) = \int h \cdot P_t 1 \, dm \leq m(h) < \infty$, $h\hat{P}_t \in L^1$, so $B(p)$ is compact. Note that

$$B(p) = \{f \in B_1^+: \ e^{pt} \, P_t f \leq f \ \text{ m-a.e. for every } t > 0\}.$$

Now let $G \in \mathcal{E}$ be chosen with $1_G \in L^1$, and let $c > 0$. Define

$$(2.5) \qquad B(p,G,c) = \bigcap_{t>0} \{f \in B(p): \ e^{pt} \int 1_G \hat{P}_t \cdot f \, dm \geq c\}$$

$$= \{f \in B(p): \ \int 1_G e^{pt} \, P_t f \, dm \geq c \ \text{ for every } t > 0\}.$$

$B(p,G,c)$ is compact.

(2.6) THEOREM. *There is a nonzero bounded eigenfunction* $f$ *with eigenvalue* $p$ *if and only if for some* $c > 0$ *and for some* $G \in \mathcal{E}$, $B(p,G,c) \neq \emptyset$.

PROOF. ($\Longrightarrow$) If $e^{pt}P_t f = f$, then $\int 1_G e^{pt} P_t f \, dm = \int 1_G f \, dm$.

($\Longleftarrow$) if $B(p,G,c) \neq \emptyset$, define a map by setting $T(g) = e^p \, P_1 g$. Let us check that $T: B(p,G,c) \to B(p,G,c)$. If $g \in B(p,G,c)$, then $T(g) \geq 0$ and $e^p \, P_1 g \leq g$ m-a.e.; so, $T(g) \in B_1^+$. Now

$$e^{pt} P_t T(g) = e^{p(t+1)} P_{t+1} g \leq g$$

m-a.e., (2.4), so $T(g) \in B(p,G,c)$. Suppose $g_n \to g$ in the weak*-topology of $B(p,G,c)$. Since $P_t(x,\cdot) \in L^1$, $P_t g_n$ converges boundedly to $P_t g$. If $h \in L^1$, the Lebesgue dominated convergence theorem gives us that

$$\int h \cdot P_t g_n \, dm \to \int h \, P_t g \, dm.$$

Since $T(rg + (1 - r)f) = rT(g) + (1 - r)T(f)$ for all f and g in $B(p,G,c)$, $0 \leq r \leq 1$, T is a continuous affine map of $B(p,G,c)$ into $B(p,G,c)$. By Lemma (2.7) below, T has a fixed point $g \in B(p,G,c)$: $T(g) = g \neq 0$. That is, $e^p P_1 g = g$ m-a.e. Since $e^{pt} P_t g \leq g$ m-a.e. and $P_t(x,\cdot) \ll m$, we obtain

$$e^{p(t+s)} P_{t+s} g(x) \leq e^{pt} P_t g(x)$$

for every x, for every t, s > 0. Let $g^* = e^p P_1 g$. Then $e^p P_1 g^* = g^*$ and $e^{pt} P_t g^* \leq g^*$. Therefore, $e^{pt} P_t g^* = g^*$ for every $t \leq 1$, and hence for every $t > 0$: $g^*$ is the desired eigenfunction. □

The following lemma is a translation of a standard result in Banach spaces [14]. In the situation where there is only one ( up to constant multiples) eigenfunction in $B(p,G,c)$, that is, there is a "unique" eigenfunction with eigenvalue p, it gives a concrete method for finding g: take the limit of the $f_n$ in (2.7) without passing to a subsequence. In general, however, there is non-uniqueness: see section 3 for several situations in which uniqueness does hold.

(2.7) LEMMA. *Let* T *be a continuous affine map of* B(p,G,c) *into itself. Then* T *has a fixed point.*

PROOF. Choose $f°$ in B(p,G,c), and set

$$f(n) = \frac{1}{n} \sum_{i=0}^{n-1} T^i(f°)$$

Since B(p,G,c) is convex, $f(n) \in B(p,G,c)$. Since B(p,G,c) is compact, there is a subsequence $(f(n_k))$ converging to f in B(p,G,c). Let $h \in L^1$, and consider

$$\int h\,(Tf - f)\,dm = \lim_{k\to\infty} \int h(Tf(n_k) - f(n_k))dm$$

$$= \lim_{k\to\infty} \int h\cdot\frac{1}{n_k}\,(T^{n_k}f° - f°)\,dm = 0. \qquad \square$$

Theorem (2.6) seems to be most useful in mildly perturbing processes which obviously have bounded eigenfunctions, in particular when the semigroup is conservative.

(2.8) EXAMPLE. Let X(t) be any process on E with $P_t 1 = 1$, and let $n_t$ be any multiplicative functional so that $E^x(n_\infty) \geq c > 0$. (For example, if X(t) is Brownian motion in $R^d$, $d \geq 3$, $n_t$ may be $\exp(-\int_0^t V(X(s))ds)$, where V is any bounded positive function with compact support.) Define the killed semigroup $Q_t g(x) = E^x[g(X(t))n_t]$. Let G be an open set in E charged by m, and let us show that $1 \in B(0,G,c/d)$ for some $d > 0$ (where B(0,G,c) is defined for $Q_t$). Since $P_t 1 = 1$, $Q_t 1 \leq 1$, so $1 \in B(0)$. Now

$$\int 1_G Q_t 1\,dm \geq \int 1_G c\,dm \geq c/d\ ,$$

so $1 \in B(0,G,c/d)$ and $Q_t$ has a positive bounded eigenfunction with eigenvalue 0. Of course, this eigenfunction is $E^x(n_\infty)$.

## 3. Uniqueness of Bounded Eigenfunctions: the Compact Case

In this section, we search for conditions guaranteeing uniqueness of bounded eigenfunctions with a given eigenvalue p. Uniqueness is a rare occurrence in general: there may be many positive bounded eigenfunctions with a given eigenvalue p. For example, if X is any process with infinite lifetime, let $T = \bigcap_{t>0} \sigma(X(s):s\geq t)$. If H is any bounded positive T-measurable random variable, then $E^x(H)$ is an eigenfunction with eigenvalue 0. So if T is nontrivial, there will be more than one eigenfunction. We are interested in simple conditions guaranteeing that there is only one (up to constant multiples, of course). If there is only one bounded eigenfunction, then (2.7) gives a concrete asymptotic method of finding it.

(3.1) PROPOSITION. *Let* E *be compact, and let* $(U^\alpha)$ *be a strong Feller resolvent on* E *so that for each* $\alpha > 0$ *and* $B \in \mathcal{E}$, *either* $U^\alpha 1_B(x) = 0$ *for all* x *in* E, *or* $U^\alpha 1_B(x) > 0$ *for all* x *in* E. *For each eigenvalue* p, *there is at most one bounded positive eigenfunction* f *(up to constant multiple).*

By strong Feller, we mean $U^\alpha: b\mathcal{E} \to C(E)$ for every $\alpha > 0$. Note that this theorem applies to diffusions with Neumann boundary conditions which are killed by a multiplicative functional.

PROOF. Suppose $f \in b\mathcal{E}^+$ and $P_t f = e^{-pt} f$. Since $U^1 f = (p + 1)^{-1} f$, we have $f \in C(E)$; so, $f \geq c > 0$ since E is compact. Let $Q_t g(x) = f(x)^{-1} P_t(fg)(x)$. Then $e^{pt} Q_t 1 = 1$; so, $R_t = e^{pt} Q_t$ is a Markov semigroup on E with strong Feller resolvent $(V^\alpha)$. From the resolvent equation, we have

$$V^0 - V^p = \sum_{k=1}^{N} p^k (V^p)^k V^p + p^{N+1} (V^p)^{N+1} V^0 .$$

If $B \in \mathcal{E}$, then either $V^p 1_B \equiv 0$ or $V^p 1_B > 0$ on E. If $V^p 1_B > 0$ on E, then $V^p 1_B \geq d > 0$ since $V^p 1_B \in C(B)$: so,

$$(V^0 - V^p) 1_B \geq \sum_{k=1}^{N} p^k (V^p)^k d \geq Nd.$$

Therefore, $V 1_B \equiv 0$ or $V 1_B \equiv \infty$. That is, $R_t$ is a recurrent semigroup, so the only invariant functions are the constants. If $R_t h = h$, then h is a constant. Translating this back to $P_t$, we have: if $P_t fh = e^{-pt} fh$ then h is a constant. □

We state a uniqueness result (for the non-compact case) which can be found in Lazer [11]. Let D be a bounded *open* set in $R^d$ so that $\partial D$ is a $C^{2+a}$ manifold for some $0 < a < 1$. Let

$$Lu = - \sum_{i,j=1}^{d} \frac{\partial}{\partial x_i}\left(a_{ij}(x) \frac{\partial u}{\partial x_j}\right) + \sum_{i=1}^{d} b_i(x) \frac{\partial u}{\partial x_i} + a_0(x) u$$

be a strongly elliptic operator in D with coefficients in $C^a(\bar{D})$. There is a positive eigenfunction $\varphi \in C^{2+a}(\bar{D})$ corresponding to the principle eigenvalue $\lambda$ so that $\varphi = 0$ on $\partial D$. If $u \in C^{2+a}(\bar{D})$ satisfies $Lu = \lambda u$ and $u = 0$ on $\partial D$, then $U = c\varphi$ for some constant c .

## 4. Yosida's Ergodic Theorem

In this section, we assume that $P_t$ is a semigroup of a transient process $X_t$ on $E_\Delta$. Suppose we know by some means that there is a non-zero positive eigenfunction f, so that $P_t f = e^{-pt} f$. Sufficient conditions for existence of such eigenfunctions have been given in preceding sections and in [19]. Can we find f or another eigenfunction by some "concrete" asymptotic procedure? Yes, in certain situations. One such has already been discussed in Section 3.

Let $E° = \{x \in E: 0 < f(x) < \infty\}$, and set

$$Q_t g(x) = e^{pt} \frac{1}{f(x)} P_t(fg)(x) .$$

Then $Q_t$ is a semigroup of a right process on $E°$ and $Q_t 1 = 1$.

(4.1) PROPOSITION. *Suppose* $Q_t$ *has a finite excessive measure* n. *If* $\bar{p}=1$ *or if* $\bar{p}=2$ *and if* $g \in L^{\bar{p}}(n)$ , *then there is a function* $g^*$ *so that* $fg^*$ *is an eigenfunction of* $P_t$ *and*

$$\left\|\frac{1}{t} \int_0^t Q_s g \, ds - g^*\right\|_{\bar{p}} \to 0 \quad \text{as} \quad t \to \infty.$$

PROOF. Let $\bar{p}=1$ or $\bar{p}=2$. If $g \in L^{\bar{p}}(n)$, then by Theorem 2 of Yosida [21 p. 333],

$$\lim_{t\to\infty} \frac{1}{t} \int_0^t Q_s g \, ds = g^*$$

exists in $L^{\bar{p}}(n)$, $Q_t(g^*) = g^*$, and $n(g) = n(g^*)$. Translating back to $P_t$, we have $e^{pt} P_t(fg^*) = fg^*$, and the desired convergence stated in the proposition above. □

REMARK. One situation in which (4.1) applies is when $E_\Delta$ satisfies (1.1) and (1.2) and $(U^\alpha)$ and $(\hat{U}^\alpha)$ are strong Feller. For then $(P_t)$ has a finite excessive eigenmeasure n and an eigenfunction f.

## 5. A Remark on a Paper of Chung and Rao

According to Newtonian electrostatics, the electrons on a negatively charged conductor will arrange themselves in a distribution which minimizes the potential energy. This distribution is called the equilibrium distribution of the conductor and agrees ( up to a constant multiple) with the equilibrium distribution obtained from Brownian motion as follows. If $L(K) = \sup\{t: X(t) \in K\}$, $(\sup \emptyset = 0)$, and if $u(x,y)$ is the Newtonian potential kernel, then

$$P^x(L(K) > 0) = \int u(x,y) \; \nu_K(dy) \quad . \tag{5.1}$$

Formula (5.1) may be used to define "equilibrium measures" for other Markov processes $X(t)$ with potential density $u(x,y)$ under fairly general conditions ([14], [6], [12]), and it is natural to ask when $\nu_K$ can be characterized as the distribution on K minimizing potential energy among all measures on K with the same mass as $\nu_K$. Generally, this holds if $u(x,y) = u(y,x)$, but is not true in the nonsymmetric case, when $u(x,y)$ may not equal $u(y,x)$. Is there an analogue in the nonsymmetric case? Energy methods have been so fruitful in the symmetric case that one hopes so.

Chung and Rao [3] have taken a first step in investigating this problem by showing that a slight modification $\pi$ of the equilibrium measure $\nu_K$ minimizes the energy of a modification $u°$ of the potential density u. The nature of the modifications needs further study, and they posed the problem of justifying the term "equilibrium distribution" as a stationary or invariant distribution in the nonsymmetric case. We recall their results and offer some elementary consequences of their work which may illumine the meaning of "equilibrium distribution." In particular, we show that $\pi$ is an invariant measure of a process obtained from X in a natural way. It is easy to see that Chung and Rao have modified the process X so that the new process and its dual have eigenfunctions with eigenvalue 1. This is in fact the key to their energy results, and we shall extend some of the energy minimization results to the unmodified process X in section 6.

In this section, we assume for simplicity that $X = (\Omega, F, F_t, X_t, \theta_t, P^x)$ is a transient Hunt process on $E_\Delta$ which is in duality with another transient Hunt process $\hat{X}$ with respect to a sigma-finite excessive measure m as described in Chapter VI of [2]. The state space E is locally compact with a countable base and with Borel field $\mathcal{E}$. The excessive functions of X and $\hat{X}$ are assumed to be lower semicontinuous. These hypotheses are somewhat different and perhaps a little stronger that those used by Chung and Rao,but this section is intended to be exploratory and provocative rather than definitive, so technicalities which arise in general discussions are undesirable here and are eschewed. We shall make further simplifying assumptions later.

Let u(x,y) be the potential density of X and $\hat{X}$. Chung and Rao start with a compact set K in E so that

$$(5.2) \qquad P^x(L(K) > 0) = 1 \text{ for all } x \text{ in } K.$$

(Recall that, in general, $\{x \in K: P^x(L(K) > 0) < 1\}$ may be semipolar.) Then,

(5.3) $$P^x(L(K) > 0) = \int u(x,y)\ v(dy)$$

for some measure v which is supported on K; v is called the equilibrium measure of K. Additionally,

(5.4) $$P^x(f(X(L(K)-));\ L(K) > 0) = \int u(x,y)\ f(y)\ v(dy)$$

for all f in $E^+$. We use the notation Ufv(x) to denote the function in (5.4). Chung and Rao use the Schauder fixed point theorem to prove a special case of (1.11);

(5.5) PROPOSITION. *There is a probability* $\pi$ *on* K *so that* $\pi(f) = \pi(Ufv)$.

(That is, they find an eigenmeasure of the kernel $V^*(x,dy) = u(x,y)v(dy)$ with eigenvalue 1. The eigenvalue turns out to be 1 because of (5.2), which also lets them use the map $\gamma \to \gamma V^*$ in the fixed point argument instead of the nonlinear map T we used in section 1.) Set

(5.6) $$\varphi(y) = \int \pi(dx)\ u(x,y).$$

Then $\varphi$ is an eigenfunction for $V^*(dx,y) = v(dx)u(x,y)$, and

(5.7) $$\pi(dy) = \varphi(y)\ v(dy)$$

by (5.5). In this special situation, if $u(x,y) > 0$, then $\pi$ is the *unique* probability satisfying $\pi(f) = \pi(Ufv)$. For if $\pi^*$ is another such probability, let $\mu = \pi - \pi^*$. If we set $\varphi^*(y) = \int \pi^*(dx)u(x,y)$, then

$$\varphi(y) - \varphi^*(y) = \int v(dx)\ [\varphi(x) - \varphi^*(x)]\ u(x,y),$$

so,

$$|\varphi(y) - \varphi^*(y)| \le \int \nu(dx)\ |\varphi(x) - \varphi^*(x)|\ u(x,y)\ .$$

But if we integrate both sides with respect to $\nu$, we get the same integral on each side. Thus $\varphi - \varphi^*$ does not change sign. Since $\nu(\varphi) = \nu(\varphi^*)$, we have $\varphi = \varphi^*$ and $\pi = \pi^*$.

We shall assume that $u(x,y) > 0$: this will also simplify our discussion of h-transforms later. Set

$$u^o(x,y) = u(x,y)/\varphi(y). \tag{5.8}$$

Then,

$$\int \pi(dx)u^o(x,y) = 1 \quad \text{for all } y \text{ in } K, \text{ and} \tag{5.9}$$

$$\int u^o(x,y)\pi(dy) = 1 \quad \text{for all } x \text{ in } K.$$

Thus 1 is an eigenfunction for the dual kernels $V(x,dy) = u^o(x,y)\pi(dy)$ and $V(dx,y) = \pi(dx)u^o(x,y)$.

Let us recall the Chung-Rao energy result. If $\lambda$ and $\mu$ are signed measures on K, define

$$\langle\lambda,\mu\rangle = \int\lambda(dx)\ u^o(x,y)\ \mu(dy), \tag{5.10}$$

whenever the integrals make sense. Set $I(\lambda) = \langle\lambda,\lambda\rangle$ , $G(\lambda) = I(\lambda) - 2\lambda(1)$, and let M be the set of signed measures on K with $I(|\lambda|) < \infty$. Let $M^1$ be the set of probability measures in M.

(5.11) PROPOSITION. *If* $I(\lambda) \ge 0$ *for every* $\lambda \in M$, *then*

(i) $G(\pi) \le G(\lambda)$ *for all* $\lambda \in M$,

(ii) $I(\pi) \le I(\lambda)$ *for all* $\lambda \in M^1$.

Proposition (5.11) contains two different forms of the statement "$\pi$ minimizes the potential energy of the density $u^o$ on K": the reader is referred to [3] for a discussion and proof. Mimicking the proof (as we do in section 6) for general eigenfunctions leads to a general energy minimization result.

Chung and Rao interpret $\pi$ as the equilibrium measure of a Markov chain and interpret $u^o$ as arising from a random time change. We propose another interpretation involving an h-transform and a time change which seems natural in light of duality and time reversal.

Let us try to find a meaning for the equations in (5.9). Notice that the function $\varphi$ defined in (5.6) is excessive for the process $\hat{X}$. This suggests that the kernel $u^o$ should be interpreted as an h-transform. The following is well-known. If $\hat{X}_t$ has a semigroup $\hat{P}_t$, let $\hat{P}^o_t$ be the semigroup on $E^\varphi = \{0 < \varphi < \infty\}$ defined by

$$\hat{P}^o_t f(x) = \frac{1}{\varphi(x)} \hat{P}_t(\varphi f)(x), \quad x \in E^\varphi . \tag{5.12}$$

Since $u > 0$, $\varphi > 0$; so, $E^\varphi = \{\varphi < \infty\}$ . Thus $E\text{-}E^o$ is polar for both X and $\hat{X}$. Let $Y_t$ be the standard process on $E^\varphi$ with semigroup $\hat{P}^o_t$. Then $X_t$ and $Y_t$ are in duality on $E^o$ with respect to the measure $\varphi(x)m(dx)$, and the potential density of the processes is $u^o$ (restricted to $E^o$).

It remains to interpret $\pi$ in (5.9): $\pi$ is the Revuz measure of dual natural additive functionals $B_{t-}$ and $\hat{B}_t$ of $X_t$ and $Y_t$. If $A_t$ is the dual predictable projection of the raw additive functional $1_{\{0 < L(K) \leq t\}}$, then it follows from (5.4) that $A_t$ has Revuz measure v. Since $\pi = \varphi v$, $B_t = \int_0^t \varphi(X_s)dA_s$ . Now, the hypothesis requiring $I(\lambda) \geq 0$ for every $\lambda$ in M is quite strong. In fact, it implies that Hunt's hypothesis (H) holds on the set K. That is, if L contained in K is semipolar for the process X, then L is polar for X (see [1] and [10])*. Since $\pi$ is supported by K, it cannot charge any semipolar set, so $B_t$ and $\hat{B}_t$ are continuous additive functionals.

Let $T_t$ and $S_t$ be the right continuous inverses of $B_t$ and $\hat{B}_t$:

$$T_t = \inf\{s: B_s > t\} \qquad S_t = \inf\{s: \hat{B}_s > t\}.$$

Once again, it is well-known that $Z_t = X(T_t)$ and $\hat{Z}_t = Y(S_t)$ are in duality with respect to $\pi(dx)$ and have potential density $u^o$ [15]. (There are certain delicate points in the last sentence which we do not dwell on: $Z_t$ has state space $\{x: P^x(T_0 = 0) = 1\}$, while $\hat{Z}_t$ may have the slightly different state space $\{x: \hat{P}^x(S_0 = 0) = 1\}$.) Let $\underline{V}^a$ and $\hat{\underline{V}}^a$ be the resolvents of $Z_t$ and $\hat{Z}_t$. By (5.9), $\underline{V}1 = \hat{\underline{V}}1 = 1$ on K. By applying formula V-5.10 in [2], we obtain $\underline{V}^a 1 = \hat{\underline{V}}^a 1 = (1 + a)^{-1}$, so the lifetimes of $Z_t$ and $\hat{Z}_t$ are exponentially distributed with parameter 1.

Let $Q_t^1$ and $\hat{Q}_t^1$ be the semigroups of $Z_t$ and $\hat{Z}_t$. Then $Q_t = e^t Q_t^1$ and $\hat{Q}_t = e^t \hat{Q}_t^1$ are Markov semigroups on K in duality with respect to $\pi$. Let $W_t$ and $\hat{W}_t$ be the processes constructed from $Q_t$ and $\hat{Q}_t$. (A sketch of the construction is given in Sec. 3 of [8].) Then $Z_t$ and $\hat{Z}_t$ are obtained by killing $W_t$ and $\hat{W}_t$ at independent exponential times. Since $Q_t 1 = \hat{Q}_t 1 = 1$, $\pi(Q_t f) = \pi(\hat{Q}_t 1 \cdot f) = \pi(f)$, so $\pi$ is an invariant measure for the processes $Z_t$ and $\hat{Z}_t$.

## 6. Minimization of Energy

In this section, we assume

---

* A correction to the proof of (3.2) in [10] will appear in the next volume of the Seminar on Stochastic Processes.

(6.1) X (resp. $\hat{X}$) is a right process with semigroup and resolvent $P_t$ and $U^\alpha$ (resp. $\hat{P}_t$ and $\hat{U}^\alpha$) on a Lusin topological state space $E_\Delta$;

(6.2) X and $\hat{X}$ are in duality with respect to a sigma-finite excessive measure m (see Chapter VI of [2]);

(6.3) There are positive functions f and $\hat{f}$ on E with $Uf = cf$ and $\hat{U}\hat{f}=\hat{c}\hat{f}$.

For simplicity, we also assume

(6.4) The potential density $u(x,y)$ is strictly positive on $E \times E$, $f < \infty$ and $\hat{f} < \infty$.

Set

$$u^o(x,y) = \frac{u(x,y)}{f(x)\hat{f}(y)}$$

$$m^o(dx) = f(x)\hat{f}(x)m(dx).$$

Then we have the analogy to (5.9):

$$(6.5) \qquad \int u^o(x,y)\, m^o(dy) = c \qquad\qquad \int m^o(dx)\, u^o(x,y) = \hat{c}.$$

In general, $c \neq \hat{c}$. The reader is referred to [18], where the product of the eigenfunctions $f\hat{f}$ plays an important role in the conditional limit theorems and time reversal discussed by Seneta.

EXAMPLE. It is worth pointing out what happens in the case where X moves uniformly to the right on $R^1$ and $\hat{X}$ moves uniformly to the left, since this behavior is reflected in the behavior of space-time processes. X and $\hat{X}$ are in duality with respect to Lebesgue measure, $e^{-px}$ is an eigenfunction of X, and $e^{qx}$ is an eigenfunction of $\hat{X}$, whenever p and q are positive. Note that p need not equal q. Note also that the product of the two eigenfunctions is never integrable.

(6.6) PROPOSITION. *If* $m^o(1) < \infty$, *then* $c = \hat{c}$.

PROOF. Integrate each equation in (6.5) with respect to $m^o$ to obtain $cm^o(1) = \hat{c}m^o(1)$. □

If $\mu$ is a measure, we let

$$U^o\mu(x) = \int u^o(x,y)\mu(dy), \qquad \mu\hat{U}^o(y) = \int \mu(dx)u^o(x,y) .$$

(6.7) COROLLARY. (i) *If* $U^o\mu \le 1$, *then* $\mu(1) \le \frac{1}{\hat{c}} m^o(1)$

(ii) *If* $\mu\hat{U}^o \le 1$, *then* $\mu(1) \le \frac{1}{c} m^o(1)$ .

PROOF. $m^o(1) \ge m^o U^o \mu = \hat{c}\mu(1)$. □

Let $\gamma$ and $\nu$ be signed measures on E. Whenever it makes sense, define

$$<\gamma,\nu> = \int \gamma(dx)U^o\nu(x)$$

$$I(\gamma) = <\gamma,\gamma>$$

$$M = \{\gamma: I(|\gamma|) < \infty, |\gamma|(1) < \infty\}$$

$$M(m^o) = \{\gamma \in M: \gamma(1) = m^o(1)\}$$

$$G(\gamma) = I(\gamma) - 2c\gamma(1)$$

*For the rest of this section, we assume that* $m^o(1) < \infty$ *so* $c = \hat{c}$. Then $I(m^o) = c\, m^o(1)$ and $G(m^o) = -c\, m^o(1) = -I(m^o)$.

(6.8) PROPOSITION. *If* $\gamma \in M$, *then* $\gamma + m^o \in M$ *and*

$$G(\gamma + m^o) = I(\gamma) - I(m^o) = I(\gamma) + G(m^o).$$

PROOF. We have

$$I(|\gamma + m^o|) \leq I(|\gamma| + m^o) =$$

$$I(|\gamma|) + I(m^o) + \langle|\gamma|, m^o\rangle + \langle m^o, |\gamma|\rangle < \infty$$

since $\gamma \in M$. So,

(6.9) $$I(\gamma + m^o) = I(\gamma) + I(m^o) + \langle\gamma, m^o\rangle + \langle m^o, \gamma\rangle$$

$$= I(\gamma) + cm^o(1) + 2c\gamma(1).$$

Now $G(\gamma + m^o) = I(\gamma) - cm^o(1) = I(\gamma) - I(m^o)$. □

(6.10) COROLLARY. *If* $\gamma \in M(m^o)$, *then* $I(\gamma) = I(m^o) + I(\gamma - m^o)$.

PROOF. Since $\gamma \in M$, $-\gamma \in M$, so $m^o - \gamma \in M$ (6.8), so $\gamma - m^o \in M$. Replacing $\gamma$ in (6.9) with $\gamma - m^o$, we obtain the result. □

We now state the following.

(P) *Positivity Principle*: If $\gamma \in M$, then $I(\gamma) \geq 0$, and $I(\gamma) = 0$ implies $\gamma = 0$.

(6.11) COROLLARY. *If* (P) *holds, then*

(i) $G(m^o) \leq G(\gamma)$ for every $\gamma \in M$;

(ii) $I(m^o) \leq I(\gamma)$ for every $\gamma \in M(m^o)$.

PROOF. (i) $G(\gamma) = G(\gamma - m^o + m^o) = I(\gamma - m^o) + G(m^o) \geq G(m^o)$ .

(ii) $I(\gamma) = I(m^o) + I(\gamma - m^o) \geq I(m^o)$ . □

Deciding when (P) holds can be very difficult. As mentioned in sec. 5, it implies Hunt's hypothesis (H).

## References


1. R.M. BLUMENTHAL and R.K. GETOOR. Dual process and potential theory. *Proc. 12th Biennial Seminar of the Canadian Math Soc.* (1970), 137-156.

2. R.M. BLUMENTHAL and R.K. GETOOR. *Markov Processes and Potential Theory*. Academic Press, New York, 1968.

3. K.L. CHUNG and K.M. RAO. Equilibrium and energy. *Prob. & Math statistics 1* (1980), 99-108.

4. K.L. CHUNG. Probabilistic approach in potential theory to the equilibrium problem. *Ann. Inst. Fourier 23* (1973), 313-322.

5. J.N. DARROCH and E. SENETA. On quasi-stationary distributions in absorbing discrete-time finite Markov chains. *J. Appl. Prob. 2* (1965), 88-100.

6. R.K. GETOOR and M.J. SHARPE. Last exit times and additive functionals. *Ann. Prob.* 1 (1973) 550-569.

7. R.K. GETOOR. *Markov Processes: Ray Processes and Right Processes* Lect. Notes in Math. 446. Springer-Verlag, Berlin-Heidelberg-New york, 1975.

8. R.K. GETOOR and J. GLOVER. Markov processes with identical excessive measures. *Math Zeitschrift*

9. R.K. GETOOR and J. GLOVER. Riesz decompositions in Markov process theory. *Trans Amer. Math Soc.* To appear.

10. J. GLOVER. Topics in probabilistic potential theory. *Seminar on stochastic Process 1982* , pp. 195-202. Birkhäuser, Boston, 1983.

11. A.C. LAZER. An extremal characterization of the principal eigenvalue of a non-self-adjoint elliptic operator. (1982) preprint.

12. P.A. MEYER. Note sur l'interprétation des mesures d'équilibre. *Seminaire de probabilités VII, pp. 210-216. Lect. Notes in Math 321.* Springer-Verlag, Berlin-Heidelberg-New York, 1971.

13. S. PORT and C. STONE. *Brownian Motion and Classical Potential Theory.* Academic Press, New York, 1978.

14. M. REED and B. SIMON. *Methods of Modern Mathematical Physics Vol 1.* Academic Press, New York, 1972.

15. D. REVUZ. Mesures associées aux fonctionnelles additives de Markov I. *Trans. Amer. Math. Soc. 148* (1970), 501-531.

16. F. RIESZ and B. SZ-NAGY. *Functional Analysis.* F. Ungar Pub. Co. New York, 1955.

17. E. SENETA. *Non-negative Matrices and Markav Chains.* Springer-Verlag Berlin-Heidelberg-New York, 1981.

18. E. SENETA. Quasi-stationary distributions and time reversion in genetics. *J. Royal Stat. Soc. Ser. B. 28* (1966), 253-277.

19. D. STROOCK. On the spectrum of Markov semigroups and the existence of invariant measures. *Lect. Notes in Math 923.* Springer-Verlag, Berlin-Heidelberg-New york, 1982.

20. D. SULLIVAN. λ-potential theory. Preprint.

21. K. YOSIDA. *Functional Analysis.* Springer-Verlag, Berlin-Heidelberg-New York 1980.

Joseph Glover
Department of Mathematics
University of Florida
Gainesville, Florida 32611

*Seminar on Stochastic Processes, 1984*
Birkhäuser, Boston, 1986

# MEAN EXIT TIMES OF MARKOV PROCESSES

by

JOSEPH GLOVER* and MING LIAO

Gray and Pinsky [3] have investigated the geometric content of the mean exit time of Brownian motion $X_t$ from a ball on a Riemannian manifold. That is, if m is the center of the geodesic ball $B_e$ of radius e, and if $T_e$ is the first time $X_t$ exits $B_e$, they obtain the asymptotic expansion of $P^m[T_e]$ as e goes to zero and identify the first three nonzero terms of the expression in terms of the geometry of the manifold. In view of the fact that $P^m[T_e]$ contains so much geometric information, it seems natural to ask to what extent $P^m[T_e]$ determines the Brownian motion itself. Could there be another process $(Y_t, Q^x)$ with $P^m[T_e] = Q^m[T_e]$ ? Without additional information, this seems difficult to determine. Restricting to exit times of balls with center m limits us severely. Our purpose in this note is to indicate that if the hypotheses are strengthened somewhat, then mean exit times do determine the process. The proof naturally appears in two parts. First, we indicate that the mean exit times determine the geometric trajectories of the process, and then we show that the speed at which the processes move along these trajectories is also determined. In fact, if two processes are time changes of one another and if the mean

* Research supported in part by NSF Grant DMS-8318204

exit times of one process dominate the mean exit times of the other process, then one process runs more slowly than the other, *path for path*. This is not true if the mean last exit times of one process dominate the mean last exit times of a time change of the process as we show in Example (5).

We stick to a fairly simple situation for ease of exposition. For Theorem (1), we fix $X = (\Omega, F, F_t, X_t, \theta_t, (P^x)_{x\in E})$ and $Y = (\Omega, G, G_t, Y_t, \theta_t, (Q^x)_{x\in E})$ two Hunt processes without traps on an LCCB state space E with Borel field $E$. For regularity hypotheses, we assume that $\hat{X}$ and $\hat{Y}$ are two more Hunt processes on E so that X and $\hat{X}$ are in duality with respect to $\xi$, Y and $\hat{Y}$ are in duality with respect to $\lambda$, and their potential densities are $u^\alpha(x,y)$ and $v^\alpha(x,y)$. (See Chapter VI of [1]. These hypotheses can be weakened a bit.) If $K \in E_\Delta$, let

$$T_K = T(K) = \inf\{t > 0: X_t \in K\}$$

$$S_K = S(K) = \inf\{t > 0: Y_t \in K\}.$$

(1) THEOREM. *If* $P^x[T_K] = Q^x[T_K]$ *for every* $x \in E$ *and for every* $K \in E_\Delta$, *then* X *and* Y *are identical in law.*

PROOF. We first prove the theorem under the additional assumption that $P^x[\zeta] < \infty$ for all x. Since $\zeta = T_{\{\Delta\}}$, $P^x[\zeta] = Q^x[\zeta] < \infty$ for all x, and we have $U1(x) = V1(x)$. Let $K \in E$ and set $K^c = E-K$. Then

$$P^x[T(E_\Delta-K)] = P^x[\zeta - (\zeta - T(E_\Delta-K))] = U1(x) - P_{K^c}U1(x) .$$

Thus $P_{K^c}U1(x) = Q_{K^c}V1(x)$, whence

$$(2) \qquad \frac{1}{U1(x)} P^x[U1(X_{T(K^c)}); T(K^c) < \zeta] = \frac{1}{V1(x)} Q^x[V1(Y_{S(K^c)}); S(K^c) < \zeta].$$

Let $(\overline{X},\overline{P}^x)$ be X conditioned by the excessive function U1(x). Then the left side of (2) is $\overline{P}^x[\,\overline{X}_t$ is in $K^c$ for some $t > 0]$. Similarly, if $(\overline{Y},\overline{Q}^x)$ is Y conditioned by V1(x), then the right side of (2) is $\overline{Q}^x[\,\overline{Y}_t$ is in $K^c$ for some $t > 0]$. That is, $\overline{X}$ and $\overline{Y}$ have the same hitting probabilities on E. Since $\overline{X}$ is obtained by conditioning X and since X has a dual $\hat{X}$, $\overline{X}$ also has a dual and a potential density $\overline{u}(x,y)$. Similarly, $\overline{Y}$ has a dual and a potential density $\overline{v}(x,y)$. It is shown in [2] under these hypotheses that there is a function f which is positive $(\xi+\lambda)$ a.e. on E so that $\overline{u}(x,y) = \overline{v}(x,y)f(y)$ $(\xi+\lambda)$-a.e. Now $\overline{u}(x,y) = U1(x)^{-1}u(x,y)$ and $\overline{v}(x,y) = V1(x)^{-1}v(x,y)$. Thus $u(x,y) = v(x,y)f(y)$ $(\xi+\lambda)$-a.e. Since $V\lambda = V1 = U1 = U\xi = Vf\xi < \infty$, $\lambda = f\xi$ by uniqueness of potentials. Therefore,

$$u(x,y)\xi(dy) = v(x,y)f(y)\xi(dy) = v(x,y)\lambda(dy),$$

so $Ug = Vg$ for every g on E. Choose a bounded function $h > 0$ on E so that $Uh = Vh$ is bounded. Let $dA_t = h(X_t)dt$, $dB_t = h(Y_t)dt$, and let $\tau_t$ and $\sigma_t$ denote the right continuous inverses of $A_t$ and $B_t$. Then $X(\tau_t)$ and $Y(\sigma_t)$ have potentials $\tilde{U}g = Ugh$ and $\tilde{V}g = Vgh$. Since $\tilde{U} = \tilde{V}$ and $\tilde{U}1 \le c$, $\tilde{U}^\alpha = \tilde{V}^\alpha$ for all $\alpha > 0$ by [1, V-5.10]. Therefore, $X(\tau_t)$ and $Y(\sigma_t)$ are identical in law, so $X_t$ and $Y_t$ are identical in law.

We now sketch the procedure necessary to extend the result to the case where $P^x[\zeta] = \infty$.

Say that a set W contained in E is a strong exit set (for X) if $W \in \mathcal{E}$, $W^c = E_\Delta - W$ is finely open, and $P^x[T(W^c)]$ is bounded in x. Let $(N_i)$ be a countable base of open sets with compact closure in E, and set

$$v_i = P^x \int e^{-t} 1_{\overline{N}_i^c}(X_t)dt \qquad W_{ij} = \overline{N}_i \cap \{v_i \ge \tfrac{1}{j}\}.$$

Blumenthal and Getoor [1,p.240] prove that each $W_{ij}$ is a strong exit set and $\bigcup_{i,j} W_{i,j} = E$. Fix i and j, and let $T = T(W_{ij}^c)$, $S = S(W_{ij}^c)$, and set

$$E_{ij} = \{x \in E:\ P^x[T] > 0\} = \{x \in E:\ Q^x[S] > 0\}$$

$$= \{x \in E:\ P^x[T > 0] = 1\} = \{x \in E:\ Q^x[S > 0] = 1\}.$$

(The last two equalities follow from the Blumenthal 0-1 law.) It is clear that

$$E_{ij} = \{x \in E:\ P^x[e^{-T}] < 1\} = \{x \in E:\ Q^x[e^{-S}] < 1\},$$

so $E_{ij}$ is finely open for both X and Y. Since $E_{ij}^c$ is finely closed, $X(T(E_{ij}^c)) \in E_{ij}^c$. Thus

$$0 = P^x\, P^{X(T(E_{ij}^c))}[T > 0] = P^x\ [T\circ\theta_{T(E_{ij}^c)} > 0],$$

and it follows that $T = T(E_{ij}^c)$. We also note that $\bigcup_{ij} E_{ij} = E$. Given x in E, one may choose an open set $N_i$ containing x so that

$$v_i(x) = P^x \int e^{-t} 1_{\overline{N_i^c}}(X_t)dt > 0$$

since X has no traps. Hence for large enough j, $x \in N_i \cap \{V_i > 1/j\}$, which is a finely open subset of $W_{ij}$ and hence contained in $E_{ij}$.

If $f \in \mathcal{E}^+$, let

$$U_{ij}f(x) = P^x \int 1_{[0,T)}(t)f(X_t)dt,$$

$$V_{ij}f(x) = Q^x \int 1_{[0,S)}(t)f(Y_t)dt$$

be the potentials of $(X,T)$ and $(Y,S)$, the processes killed the first time they exit $W_{ij}$ (or $E_{ij}$). We have $U_{ij}1 = V_{ij}1$ on $E_{ij}$.

It follows as in the first case that $U_{ij}g = V_{ij}g$ for every g on $E_{ij}$, so $(X,T)$ and $(Y,S)$ are identical in law on $E_{ij}$. It remains to piece together all of the results on the various $E_{ij}$. This is a long procedure carried out in V-5 of Blumenthal and Getoor [1], to which we refer the reader. □

The result above shows that mean exit times specify the speed of the process. Now we go on to connect domination of mean exit times with domination of velocities.

No extra effort is required to allow $X = (\Omega, F, F_t, X_t, \theta_t, (P^x)_{x\in E})$ and $Y = (\Omega, G, G_t, Y_t, \theta_t, (Q^x)_{x\in E})$ to be two right processes without traps on $(E,\mathcal{E})$. Let $A_t$ be a continuous additive functional of $X_t$ which is strictly increasing on $[0,\zeta)$, where $\zeta = \inf\{t: X_t = \Delta\}$. We set $\tau_t = \inf\{s > 0: A_s > t\}$, and we assume that $(X(\tau_t),P^x)$ and $(Y_t,Q^x)$ have the same law. It follows that $Y_t$ can also be time changed to have the same law as $X_t$, so we are assuming that X and Y have the same geometric trajectories. Once again, if $K \in \mathcal{E}$, let $T_K$ and $S_K$ be defined as above.

(4) THEOREM. *Under the hypotheses in the paragraph above, if* $P^x[T_K] \le Q^x[S_K]$ *for all* K *contained in* $E \cup \{\Delta\}$, *then* $\tau_t \le t$ a.s.

PROOF. Since $(X(\tau_t),P^x)$ is equivalent in law to $(Y_t,Q^x)$, it follows that

$$P^x[A(T_K \wedge \zeta)] = Q^x[S_K \wedge \zeta].$$

Note that $T_K \wedge \zeta = T_L$ and $S_K \wedge \zeta = S_L$, where $L = K \cup \{\Delta\}$. Combining this fact with the hypothesis, we get $P^x[T_K \wedge \zeta] \le P^x[A(T_K \wedge \zeta)]$ for every $K \in \mathcal{E}$. Define an additive functional $B_t = t \wedge \zeta$, so we have $P^x[B(T_K)] \le P^x[A(T_K)]$. Let $D_t = B_t + A_t$, and let $\sigma_t$ be the right continuous inverse of $D_t$. Choose $(W_{ij})$ as in Theorem (1) for the process $X(\sigma_t)$ so that $P^x[R]$ is bounded, where $R = \inf\{t: X(\sigma_t) \in W_{ij}^c\}$. Since $R = D(T(W_{ij}^c))$, $v(x) = P^x[B(T(W_{ij}^c))]$ and $u(x) = P^x[A(T(W_{ij}^c))]$ are bounded. Now u and v are excessive for X killed at $T(W_{ij}^c)$, and $w = u - v \ge 0$. Moreover, if $K \subset E_{ij}$ and $T = T(W_{ij}^c)$,

$$P_K w = P^x[(A_T - A_{T_K}) - (B_T - B_{T_K})]$$

$$= u - v - P^x[A_{T_K} - B_{T_K}] \le w .$$

Since w is finely continuous on $E_{ij}$ for $(X,T)$, $w = E^x[C_T]$ for some $(X,T)$-continuous additive functional $C_t$. Hence $t < A_t$ for $t < T(E_{ij}^c) = T(W_{ij}^c)$. Now let $R = \inf\{s > 0: s > A_s\}$. Since $\cup_{i,j} E_{ij} = E$, $R > 0$ a.s. Since $A_{R+t} = R + A_t \circ \theta_R$ and $R > 0$,

$$P^x[\text{there is an } e>0 \text{ such that } A_{R+t} < R+t \text{ for all } t<e]$$

$$= P^x[\text{there is an } e>0 \text{ such that } A_t \circ \theta_R < t \text{ for all } t<e]$$

$$= P^x[P^{X(R)}[\text{there is an } e>0 \text{ such that } A_t < t \text{ for all } t<e]] = 0.$$

Thus $A_{R+t} \ge t + R$ a.s. on $\{t < \zeta\}$, so $R \ge \zeta$ a.s. Thus $t \le A_t$ for every $t \le \zeta$, so $\tau_t \le t$. □

Now we show that the conclusion of the theorem above is false if we replace $P^x[T_K] \le Q^x[T_K]$ with $P^x[L_K] \le Q^x[L_K]$, where $L_K$ is the last exit time from K.

(5) EXAMPLE. Let $E = [0,2)$, $\Delta = \{2\}$, and let $X_t$ be a Hunt process on $E_\Delta$ which may be described as follows. While X is in $[0,1)$, it moves to the right with unit speed. When it enters $[1,2)$, an exponential-(1) alarm clock starts ticking. If it does not ring before the process hits $\Delta = \{2\}$, the process dies there. If it rings at time T before the process hits $\Delta$, it jumps to $\{0\}$, where the whole motion begins afresh.

Let $A_t$ be a continuous additive functional defined by setting $dA_t = f(X_t)dt$, where

$$f(x) = \begin{cases} 1/2 & \text{if } x \in [0,1), \\ N & \text{if } x \in [1,2), \end{cases}$$

where $N \ge 1$ is a large number to be determined later. It is clear that $A_t$ is strictly increasing. If $X_0 \in [0,1)$, then $t > A_t$ for small t, while if $X_0 \in [1,2)$, $t < A_t$ for small t. Let $\tau_t$ be the right continuous inverse of $A_t$, and let $Y_t = X(\tau_t)$. We now check that if $K \subset [0,2)$, $L = L_K = \sup\{t: X_t \in K\}$, and $M = M_K = \sup\{t: Y_t \in K\}$, then $P^x[L] \le P^x[M]$. This is equivalent to checking that

$$P^x[L] \le P^x[A_L]. \tag{6}$$

Define $T_1 = T$ and set $T_n = T_{n-1} + T\circ\theta_{T_{n-1}}$. It is clear that $X_{T(n)} = 0$ on $\{T_n < \zeta\}$. Fix x in $[0,2)$, and let $b = \sup\{(y - x) \vee 0: y \in K\}$. Put $\Lambda_n = \bigcap_{k\le n}\{T_k < \zeta\}$. By the strong Markov property,

$$P^x[\Lambda_n] = P^x[T < \zeta]P^0[T < \zeta]^{n-1} = P^x[T < \zeta](1 - e^{-1})^{n-1} .$$

Thus,

$$P^x[L] = P^x[L; T \geq \zeta] + \sum_{n=1}^{\infty} P^x[L; \Lambda_n - \Lambda_{n+1}]$$

$$\leq P^x[L; T \geq \zeta] + \sum_{n=1}^{\infty} 4n\, P^x[\Lambda_n - \Lambda_{n+1}]$$

$$= P^x[L; T \geq \zeta] + 4e\, P^x[T < \zeta] .$$

On the other hand,

$$P^x[A_L] \geq P^x[A_L; T \geq \zeta] + P^x[A(T_2); T < \zeta, T_2 < \zeta]$$

$$\geq P^x[A_L; T \geq \zeta] + P^x[T < \zeta]P^0[A_T; T < \zeta]$$

$$\geq P^x[A_L; T \geq \zeta] + P^x[T < \zeta] \int_0^1 e^{-s} Ns\, ds.$$

Thus, for x in [0,1),

$$P^x[L] \leq 2e^{-1} + 4e(1 - e^{-1}) \leq (1 - e^{-1})N \int_0^1 se^{-s} ds \leq P^x[A_L]$$

for N sufficiently large. For x in [1,2),

$$P^x[L] \leq be^{x-2} + 4e(1 - e^{x-2})$$

$$\leq Nbe^{x-2} + (1 - e^{x-2})N \int_0^1 se^{-s} ds \leq P^x[A_L]$$

for N sufficiently large. Therefore, if N is chosen sufficiently large,

$P^x[L] \leq P^x[A_L]$ for all x in [0,2).

REMARK. The process constructed above is a Hunt process; i.e. it is quasi-left continuous. A simpler example can be obtained as follows by dropping the quasi-left continuity. Let $X_t$ be uniform motion to the right on [0,2). Upon reaching 2, the process jumps to a cemetery $\Delta$ with probability 1/2 and jumps to 0 with probability 1/2.

## References

1. R.M. BLUMENTHAL and R.K. GETOOR. *Markov Processes and Potential Theory*. Academic Press, New York, 1968.

2. J. GLOVER. Markov processes with identical hitting probabilities. *Trans. Amer. Math. Soc. 275* (1983), 131-141.

3. A. GRAY and M. PINSKY. The mean exit time from a small geodesic ball in a Riemannian manifold." *Bull. Sc. Math. 107* (1983), 345-370.

Joseph Glover
Department of Mathematics
University of Florida
Gainesville, Fla. 32611

Ming Liao
Department of Mathematics
Stanford University
Stanford, Ca. 94305

*Seminar on Stochastic Processes, 1984*
Birkhäuser, Boston, 1986

# ON STRICT-SENSE FORMS OF THE HIDA-CRAMER REPRESENTATION*

by

FRANK B. KNIGHT

## §1. The Gaussian Case

This paper is an attempt to delineate more clearly than before the role of the standard Brownian motion and Poisson processes in generating general stochastic processes $(\Omega, \mathcal{Z}_t, X_t, P)$. In discrete time, the analog would be to use a sequence of independent Bernoulli random walks. This is a very different setting, and one about which we have nothing to contribute. Evidently such a sequence does not go far toward generating a general discrete parameter process, at least in the sense we have in mind here. The situation in continuous time, however, is antithetical. One can obtain representation theorems of considerable generality, which to some extent crystalize an important aspect of all continuous time processes. We consider this as the aspect which involves randomness of time *without* randomness of place. In some sense, there is a natural dichotomy of the two kinds of randomness, and our aim is to isolate and study the case in which the role of randomness of place can be eliminated. A basic tool in our investigation is the "prediction process" $Z_t$ of [11], but it is no more than that. Theoretically, one could attempt to define respresentations of $Z_t$ itself, analogous to those

*Research supported in part by N.S.F. Grant 83-03305.

obtained below but valid for all P. In the present paper, however, each P is treated separately.

A natural place to begin is with the Gaussian case, or more generally (in the language of [5, p. 77]) with the wide-sense aspect of the representation problem. Here the key result was given by T. Hida [9, 1960] and H. Cramer [1, 1961] for processes, although it was known much earlier in the setting of Hilbert space multiplicity theory. We will follow mainly [9] in our statement. The essential part of the result can be proved quite easily by probabilistic argument, as in [14] for example.

Let $(\Omega,\mathcal{Z},P)$ be a complete probability space, and on it let $X_t$, $t \geq 0$, be a real-valued $L^2$-process (i.e., $EX_t^2 < \infty$). Denote by $H(t)$ the Hilbert space closure of the set $\{X_s,\ s \leq t\}$, with $H(t+) = \bigcap_{\varepsilon>0} H(t+\varepsilon)$. We make two assumptions:

a) $H(0+) = 0$ (the trivial Hilbert space), and

b) $H(\infty)$ is separable. Then we have the following representation.

THEOREM 1.1. *There is a unique index* $E(t) \leq \infty$, *called the index of multiplicity, which is non-decreasing in* $t \geq 0$, *such that there is a sequence* $G_k(t) \subset H(t+)$ *of orthogonal processes with orthogonal increments, right-continuous in* $H(\infty)$, *with* $G_k(0) = 0$ *and* $G_k(t) = 0$ *for* $k > E(t)$, *whose squared norms* $V_k(t) = EG_k^2(t)$ *satisfy* $dV_1 \gg dV_2 \gg \cdots$ *in the sense of absolute continuity of measures, and such that for every* $X \in H(t+)$ *there are measurable functions* $F_k(s)$, $k \leq E(t)$, *such that*

$$X = \sum_{k<E(t)+1} \int_0^t F_k(s)\,dG_k(s), \tag{1.1}$$

*where the integrals and series converge in the usual* $L^2$*-sense.*

REMARK. This is essentially Theorem 1.5 of [9], with three changes of notation. We use $E(t)$ for the multiplicity index instead of for the entire projection operator onto $H(t+)$. Secondly, we represent all of $H(t+)$, instead of only $X_s$, $s \le t$. Thirdly, we do not assume that $X_t$ is Gaussian with mean 0. Only the second change may require comment. We have $X \in H(t+\varepsilon)$, $\varepsilon > 0$, whence it suffices to see that any $X \in H(t+\varepsilon)$ has a representation (1.1) with $t+\varepsilon$ in place of $t$, and then note that $\int_t^{t+\varepsilon} F_k(s)dG_k(s)$ is orthogonal to $X$ for each $k$ if $X \in H(t+)$. The former fact is obtained by passing to limits in $L^2$ of expressions $\sum_{j=1}^n c_j X_{s_j}$, $s_j \le t+\varepsilon$, using the familiar isomorphism of $H(t+\varepsilon)$ and the direct sum $\sum_{k<E(t+\varepsilon)+1} L^2(dV_k(s), s \le t+\varepsilon)$.

We have stated this result only by way of introduction. What we are really interested in studying is its strict sense analog. We recall from [5] that a strict sense result is a stronger property which holds when a wide sense result is specialized to the case of Gaussian $X_t$ with mean 0. In the present case there are various such analogs, but one which serves to introduce the main idea is given by Davis and Varaiya [3] albeit (as will be seen) they overstate the analogy. Instead of a) and b), we suppose to begin with a filtration $\mathcal{Z}_t^\circ$, augmented in the usual way to a right-continuous filtration $\mathcal{Z}_t \supset \mathcal{Z}_{t+}^\circ$, containing all P-null sets. We assume

a') $\mathcal{Z}_{0+}^\circ$ contains only sets equivalent to $\phi$ or $\Omega$,

b') $L^2(\Omega,\mathcal{Z}_\infty,P)$ is separable (it suffices for this that each $\mathcal{Z}_t^\circ$ be countably generated). Then letting $L_0^2(\mathcal{Z}_t)$ denote $\{X \in L^2(\mathcal{Z}_t): EX = 0\}$, we have

THEOREM 1.2. *There is a unique index* $N(t) \le \infty$, *called the* 2-*dimension, which is non-decreasing in* $t \ge 0$, *such that if* $\mathfrak{M}_0^2$ *(resp.* $\mathfrak{M}_0^2(t)$*) is the space of right-continuous* $\mathcal{Z}_s$*-martingales* $M(s)$, $M(0)=0$, *which are* $L^2$*-bounded (resp. for* $s \le t$*) there are* $M_k \in \mathfrak{M}_0^2$,

$1 \le k < N(\infty)+1$ *(resp.* $< N(t)+1$*), orthogonal in the martingale sense* ($M_j M_k$ *is a martingale for* $j \neq k$*), whose quadratic variation processes* $\langle M_k \rangle_t$ *satisfy* $d\langle M_1 \rangle >> d\langle M_2 \rangle >> \cdots$*, with* $M_k(t) = \langle M_k \rangle_t = 0$ *for* $k > N(t)$*, and such that for every* $X \in L_0^2(\mathcal{Z}_t)$ *there are* $\mathcal{Z}_t$*-previsible processes* $h_k$*,* $k < N(t)+1$*, such that*

$$(1.2) \qquad X = \sum_{k<N(t)+1} \int_0^t h_k(s)dM_k(s),$$

*where the integral is the usual previsible stochastic integral* [4, Chap. VIII].

REMARK. This follows from Theorems 1 and 2 of [3] when we represent the martingale $m_s = E(X|\mathcal{Z}_s)$, $0 \le s \le t$.

Comparing the representations (1.1) and (1.2), it is plausible to imagine that if we specialize to Gaussian $X_t$ with mean 0 in (1.1), then (1.2) should follow as well, using $N(t) = E(t)$ and $M_k(t) = G_k(t)$. In other words, (1.2) should be a strict sense analog of (1.1). It is even claimed in Remark 1 of [3] that this is the case, and that it is implied by a certain result in Kunita and Watanabe [13]. However, we shall see that this is not the case in general, and even when it is true it does not seem to follow from that result except under further restrictions.

Our first goal is therefore to establish the necessary and sufficient condition for (1.2) to follow from (1.1) in the mean 0, Gaussian case, and to see when $E(t) = N(t)$ holds.

THEOREM 1.3. *In the Gaussian case,* (1.2) *holds with* $M_k = G_k$ *and* $N(t) = E(t)$ *(from* (1.1)*) if and only if all the* $V_k(t)$ *are continuous. The equality* $N(t) = E(t)$ *holds at time* $t$*, without requiring* $M_k = G_k$*, if* *and only if either the* $V_k(s)$ *are all continuous in* $(0,t]$ *or* $E(t)=\infty$.

Before giving the proof, here is an immediate

COROLLARY 1.4. *Let* $(G_k(t),\ k < K+1)$, $K \le \infty$, *be independent, mean* 0 *Gaussian processes with independent increments with variances* $dV_1 \gg dV_2 \gg \cdots$ *as in* (1.1). *Then every* $X \in L_0^2(\mathcal{Z}_\infty^\circ)$, *where* $\mathcal{Z}_\infty^\circ$ *is the generated* σ-*field, has an expression* (1.2) *(with* $N(t)=K$*) if and only if the* $V_k(t)$ *are all continuous.*

Indeed, as shown by H. Cramer [2], every such family $G_k(t)$ can be determined as in (1.1) from a Gaussian process $X_t$ (actually, $X_t$ may even be harmonizable). Hence the Corollary reduces to the Theorem..

REMARK. This generalizes the result from Kunita and Watanabe [13] cited above (which is closely related to an earlier representation of K. Ito), to the effect that such a representation (1.2) holds if $K < \infty$ and each $G_k$ is a Brownian motion. Conversely, if $K < \infty$ and each $dV_k$ is equivalent to Lebesgue measure, then the Corollary follows trivially from that result.

Turning to the Theorem, note first that since the $G_k$ are orthogonal in $\mathfrak{M}_0^2$, and $\langle G_k \rangle_t = V_k(t)$ is deterministic in the Gaussian case, we must have $E(t) = N(t)$ whenever (1.2) is valid with $G_k$ and $E(t)$ in place of $M_k$ and $N(t)$. This follows by the uniqueness of $N(t)$ (Theorem 2 of [3]) since any two such representations must have the same number of elements.

Suppose, now, that some $V_k(t)$ has a discontinuity, say $\Delta V_k(t_0) > 0$, and let us observe that this implies $N(t_0) = \infty$. Indeed, (1.2) at $t = t_0$ then becomes

$$X = \sum_{k<N(t_0)+1} \left(\int_0^{t_0-} h_k(s)dG_k(s) + h_k(t_0)\Delta G_k(t_0)\right)$$

where the integrals, and also $h_k(t_0)$, are all in $\mathcal{Z}(t_0-)$. But $\Delta G_k(t_0)$ is independent of $\mathcal{Z}(t_0-)$. Thus if $\Delta V_k(t_0) > 0$, the sequence $X_n = \sin(n\pi\phi(\Delta G_k(t_0))$, where $\phi$ is the distribution function of $\Delta G_k(t_0)$, is easily seen to be orthogonal, and independent of both $\mathcal{Z}(t_0-)$ and $\Delta G_j(t_0)$, $j \neq k$. To represent this sequence in the form (1.2) is thus impossible using $M_k = G_k$, and any other choice of $M_k$ requires a whole sequence of orthogonal discontinuities at $t = t_0$ to write $X_n = \sum_{k<N(t_0)+1} h_{k,n}(t_0)\Delta M_k(t_0)$ for all $n$. Indeed, since $X_n$ is independent of $\mathcal{Z}(t_0-)$, $E(X_nX_m) = E(E(X_nX_m|\mathcal{Z}(t_0-))) = \sum_{k<N(t_0)+1}[h_{k,n}(t_0)h_{k,m}(t_0) E(\Delta\langle M_k\rangle_{t_0}|\mathcal{Z}(t_0-))]$. Thus with probability one the sequences $(h_{k,n}(t_0))$, $1 \leq n$, are orthogonal in $n$ with respect to the weights $E(\Delta\langle M_k\rangle_{t_0}|\mathcal{Z}(t_0-))$, and of strictly positive norm, which is impossible unless $N(t_0) = \infty$.

To complete the proof, it now suffices to show that if the $V_k(t)$ are continuous then (1.2) applies with $M_k = G_k$ and $N(t) = E(t)$. The method of [13] depended on the homogeneity of Brownian motion, which is not available for the $G_k$. Instead, we will substitute the author's "prediction process" from [11, Essay I] which is always homogeneous.

LEMMA 1.5. *If the* $V_k(t)$ *are continuous then all* $M(t) \in \mathfrak{M}_0^2$ *are continuous.*

PROOF. Continuity of $M(t)$ is a property which depends only on the joint distributions at rational times, as is easily seen. Moreover, if the $V_k(t)$ are continuous then the $G_k(t)$, chosen right-continuous, are also continuous. Since they generate $H(t+)$ by (1.1), they also generate $\mathcal{Z}_t$. Indeed, by independence of increments the $\sigma$-field $\sigma\{G_k(s), s \leq t\}$ is independent of

$$\sigma\{G_k(t+s) - G_k(t),\ 0 \le s\} = \bigvee_{\varepsilon>0} \sigma\{G_k(t+\varepsilon+s) - G_k(t+\varepsilon),\ 0 < s\}.$$

Therefore the former is equivalent to $\bigcap_{\varepsilon>0} \sigma\{G_k(s),\ s \le t+\varepsilon\}$, since either, together with the independent $\sigma\{G_k(t+s) - G_k(t),\ 0 < s\}$, generates all of $\mathcal{Z}^\circ_\infty$. Now to prove the lemma, it suffices to reintroduce $P$ as $P_c$ on the canonical space $\Omega_c = \{w_k(t),\ 1 \le k$, continuous in $t \ge 0\}$, $\mathcal{Z}^\circ_c(t) = \sigma\{w_k(s),\ s \le t\}$, $P_c\{\bigcap_k (w_k(t_k) \in B_k)\} = P\{\bigcap_k (G_k(t_k) \in B_k)\}$ in the usual way.

On $(\Omega_c, \mathcal{Z}^\circ_c, P_c)$, let $Z_t(S)$, $S \in \mathcal{Z}^\circ_c$ (= the usual canonical $\sigma$-field) denote the prediction process of $P_c$, defined by $Z_t(S) = P_c(\theta_t^{-1} S \mid \mathcal{Z}^\circ_c(t+))$, and $Z_t(S)$ is $\mathcal{Z}_c(t)$-optional for each $S$. Then $Z_t$ is also a homogeneous Markov process, and in a suitable topology of probability measures on $\mathcal{Z}^\circ_c$ it is a realization of a Borel right process, in the sense of Meyer and Walsh [20] and Getoor [7]. This is directly from Theorem 1.17, Essay I, of [11]. Moreover, we know that $\{Z_s,\ s \le t\}$ generates $\mathcal{Z}^\circ_c(t+)$ up to $P_c$-null sets (Theorem 1.9, Essay I, of [11]). As in [ibid, Proposition 1.3], let $H$ be the metric space of probabilities on $\mathcal{Z}^\circ_c$, let $H_0^+$ be the Ray-Knight compactification of its non-branching points [ibid, p. 25], and finally let $R^Z_\lambda$ denote the resolvent $Z_t$ on $H_0^+$ with Borel field $\mathcal{H}$. According to a basic result of Kunita and Watanabe [13], which extends without alteration to any realization of a right process (for example, we can simply repeat the proof from P. A. Meyer [17, Théorème 6, p. 131] if we delete the $u \cdot \bar{X}_0$ appearing on p. 132, line 6), in order for every $M(t) \in \mathcal{M}^2_0$ to be continuous it is necessary and sufficient that every $M(t)$ of the form

$$M(t) = R^Z_\lambda f(Z_t) - R^Z_\lambda f(Z_0) + \int_0^t (f(Z_s) - R^Z_\lambda f(Z_s))ds$$

be continuous, for $f$ continuous on $H_0^+$. This is the same as continuity of $R^Z_\lambda f(Z_t)$, where $R^Z_\lambda f$ are continuous on $H_0^+$ (in the Ray topology) and separate points. Accordingly, by the Stone-Weierstrass Theorem this also is the same as continuity of $Z_t$ itself in the Ray topology.

On the other hand, continuity of $Z_t$ can be defined without reference to any particular topology. The "left-limit process" $Z_{t-} \in H$ is defined uniquely by

$$Z_{t-}(S) = P(\theta_t^{-1} S | \mathcal{Z}_c^\circ(t-)) \tag{1.3}$$

and $\mathcal{Z}_c(t)$-previsibility. Then it follows by the previsible section theorem that the left-limit process of $R_\lambda^Z f(Z_t)$ is given by $R_\lambda^Z f(Z_{t-})$ for all t, hence $Z_t$ is continuous if and only if $Z_{t-} = Z_t$ for all $t > 0$, $P_c$-a.s. To show this here, it suffices to show that $E^{Z_{t-}} f_n = E^{Z_t} f_n$ for any countable family $f_n$ such that $E^z f_n$ determines the measure z uniquely.

One such family $f_n$ can be written in the form

$$\{f_n = g_k(w_1(r_1), w_2(r_2), \ldots, w_m(r_m)\}$$

where $g_k(x_1,\ldots,x_m)$ varies over a countable dense set in $C_0(R^m)$, for each m, and $(r_1,\ldots,r_m)$ vary over the positive rationals. For such $f_n$, recalling that $w_j(r_j+t) - w_j(t)$ is independent of $\mathcal{Z}_c^\circ(t+)$ and Gaussian, with variance $V_j(r_j+t) - V_j(t)$ continuous in t, we see easily that

$$\begin{aligned} E^{Z_t} f_n = E(g_k((w_1(t+r_1) - w_1(t)) + w_1(t), \ldots \\ (w_m(t+r_m) - w_m(t)) + w_m(t)) | w_1(t), \ldots, w_m(t)), \end{aligned}$$

which varies continuously in t as required.

Continuing with the proof of Theorem 1.3 we assume for purposes of reductio ad absurdum that there is an $X \in L_0^2(\mathcal{Z}_\infty)$ for which (1.2) does not hold with $t = \infty$ and $M_k = G_k$. Then $X_t = E(X|\mathcal{Z}_t) \in \mathfrak{M}_0^2$ fails to be representable as in (1.2) for some $t$, which we may suppose is $t=1$. By a familiar orthogonalization procedure of Kunita and Watanabe [13] we can write

$$X_t = \Big(\sum_{k<E(1)+1} \int_0^t \langle X,G_k\rangle_s \, dG_k(s)\Big) + Y_t, \qquad 0 \le t \le 1,$$

where $Y_t \in \mathfrak{M}_0^2$ is orthogonal to each $G_k$, and $EY_1^2 > 0$. By Lemma 1.5, $Y_t$ has continuous paths, P-a.s.

To explain the next step most clearly, it is convenient to again make a change of the probability space (in Section 2 we shall see how this could be formally avoided, but here it also serves as an introduction). Let $W_0,W_1,\dots,W_k$ $(k < E(1)+1)$ be independent standard Wiener processes on a separate probability space, and let $(\Omega^*,\mathcal{Z}^*,P^*)$ denote the product probability space of $(\Omega,\mathcal{Z},P)$ and this auxiliary space. On $(\Omega^*,\mathcal{Z}^*,P^*)$ we define, in the obvious way,

$$G_k^*(t) = \begin{cases} G_k(t) & \text{if } t \le 1 \\ G_k(1) + W_k(t-1) & \text{if } t > 1 \end{cases} \qquad \text{and also}$$

$$Y^*(t) = \begin{cases} Y(t) & \text{if } t \le 1 \\ Y(1) + W_0(t-1) & \text{if } t > 1. \end{cases}$$

Now it is trivial to see that $G_k^*$ and $Y^*$ are orthogonal, continuous martingales relative to their generated $\sigma$-fields. The reason for the construction is to obtain that $\lim_{t\to\infty} \langle G_k^*\rangle_t = \lim_{t\to\infty} \langle Y^*\rangle_t = \infty$, and this is also obvious. Accordingly, let $\tau_k^*(t) = \inf\{s:\langle G_k^*\rangle_s \ge t\}$ and $\tau_0^*(t) = \inf\{s:\langle Y^*\rangle_s \ge t\}$. Then it follows immediately by a theorem of the author [10] that the processes $(Y^*(\tau_0^*(t)),G_k^*(\tau_k^*(t));\ k < E(1)+1))$ are independent ordinary Brownian motions ($\equiv$ Wiener processes).

We now reach a contradiction as follows. Since the $G_k(t)$, $t \le 1$, generate $\mathcal{Z}_1$, so the $G_k^*(t)$, $t \le 1$, generate a copy $\mathcal{Z}_1^*$ of $\mathcal{Z}_1$ in $(\Omega^*,\mathcal{Z}^*,P^*)$, and since $\tau_k^*(t) = \inf\{s: V_k(s) \ge t\}$ for $t \le V_k(1)$, which is a non-random function, it is clear that $\mathcal{Z}_1^* \subset \sigma\{G_k^*(\tau_k^*(\cdot)),$ $k < E(1) + 1\}$. Moreover, it is not hard to see that $Y^*(\tau_0^*(t \wedge \langle Y^* \rangle_1))$ is measurable over $\mathcal{Z}_1^*$ for all $t \geq 0$, because $Y_t^* \equiv Y_t$ for $t \leq 1$ and $\tau_0^*(t \wedge \langle Y^* \rangle_1) \leq 1$. Thus it would follow that $Y^*(\tau_0^*(t \wedge \langle Y^* \rangle_1))$ is measurable over $\sigma(G_k^*(\tau_k^*(\cdot)), k < E(1) + 1)$, whereas $Y^*(\tau_0^*(\cdot))$ is entirely independent of the latter. This is absurd unless $\langle Y^* \rangle_1 = 0$, P-a.s. (indeed, since $\langle Y^* \rangle_1 \in \mathcal{Z}_1^*$, if $\mathcal{Z}_1^*$ were given the process $Y^*(\tau_0^*(t \wedge \langle Y^* \rangle_1))$ would be both a Wiener process up to $t = \langle Y^* \rangle_1$ and a fixed function of $t$). By our original hypothesis $EY_1^2 > 0$, hence we cannot have $\langle Y^* \rangle_1 \equiv 0$ and the proof is complete.

## 2. Strict-Sense Extensions

According to the description of [5, p. 77], a strict-sense property is one which, when it is specialized to the Gaussian (mean 0, complex case), is implied by a weaker property stated in terms of variances and covariances. Here we have specialized to the real case, but that is (probably) unimportant. A more important limitation is that in order to interpret the 2-dimension $N(t)$ as a strict-sense index of multiplicity we must assume continuity of the $V_k(t)$. The question immediately arises of how to express this in the non-Gaussian case. There is not a unique answer to this question, and the stricter our interpretation is, the more nearly will our results resemble those of the Gaussian case. We will consider four interpretations, in order of decreasing generality. These are:

<u>H1</u>. $E\langle M_k \rangle_t$ is continuous, all $k$.

<u>H2</u>. $\langle M_k \rangle_t$ is continuous, all $k$.

Equivalently, the family $\mathcal{Z}_t$ is quasi-left-continuous.

<u>H3</u>. There are no pure-jump martingales in the sense of Le Jan [15]. Equivalently, for all $\mathcal{Z}_t$-optional $T$, $\mathcal{Z}_{T-} = \mathcal{Z}_T$ [15, 3. a), p. 220].

<u>H4</u>. All elements $M(t) \in \mathfrak{M}_0^2$ are a.s. continuous.

We will treat these in the order H1, H2, H4, H3. Our main result concerns H3 which seems to be in a sense the most appropriate level of generality. Our reason for not immediately accepting H1 is, that in the Gaussian case, (1.2) has another form which gives rise to a more limited, but also more useful, generalization. Namely, we have a representation in terms of independent Brownian motions $B_k$ of the form

$$(2.1) \qquad X = \sum_{k<N(t)+1} \int_0^\infty h_k(\tau_k(u-))dB_k(u \wedge V_k(t))$$

where $\tau_k(t) = \inf\{s: V_k(s) \geq t\}$, and $B_k(u) = M_k(\tau_k(u))$, $u < \lim_{t\to\infty} V_k(t)$. This follows directly by change of variables since it is easily seen that, for $s_1 \leq t$,

$$\int_0^t I_{[0,s_1)}(s)dM_k(s) = \int_0^{V_k(t)} I_{[0,s_1)}(\tau_k(u-))dB_k(u).$$

The essential feature of (2.1) is, of course, that $(B_k(u), k < N(\infty)+1)$ are independent Brownian motion processes in their time spans $(0, \lim_{t\to\infty} V_k(t)]$. It is this kind of a representation which we are interested in extending to the non-Gaussian case.

It is not hard to see that nothing like (2.1) holds under H1 or H2. Indeed, let $X$ be any non-trivial random variable with $EX^2 < \infty$, $EX = 0$, and let $e$ be an exponential random variable independent of $X$. Then the process $X_t = XI_{[e,\infty)}(t)$ generates a filtration satisfying H1 and H2 since $\langle X\rangle_t = (e \wedge t)EX^2$. If our aim is a reduction to stochastic integrals of Brownian motions (or Poisson processes), such a process $X_t$ cannot be allowed because, at the optional time $T = e$, we do not have

$X_T \in \mathcal{Z}_{T-}$. In short, there is randomness of place at time T, as well as randomness of time alone. We will show below that, if we allow both independent Brownian motions and Poisson processes as integrators in (2.1), such a representation is possible only under H3. Consequently, H3 seems to best epitomize the randomness of time alone, which (we would argue) is the essential meaning of continuity of $V_k(t)$ in the Gaussian case. To realize that this is a way of viewing Brownian filtrations, one should recollect that a Brownian motion $B(t)$ can be written as a limit of scaled Poisson processes $n^{-\frac{1}{2}}P_n(t) - n^{\frac{1}{2}}t$, where $P_n$ has parameter $\lambda = n$. Evidently the randomness of such processes reduces to that of their times of discontinuity alone.

Before treating the more general H3, let us consider briefly H4.

THEOREM 2.1. *Under H4, representation* (2.1) *remains valid in the non-Gaussian case, provided that we replace* $V_k(t)$ *by* $\langle M_k \rangle_t$ *and permit* $h_k(u)$ *to be any* $\mathcal{Z}_u$*-previsible process with* $E\int_0^t h_k^2(u)d\langle M_k\rangle_u < \infty$, $k < N(t) + 1$.

PROOF. In a formal sense, this is just an immediate application of a known change-of-variables formula for stochastic integration (see, for example, [16, p. 390, Prop. 3 and Remarque (10)], where our $B_k(u \wedge \langle M_k \rangle_t)$ coincides with $\mathcal{J} M_k(u \wedge t) = M_k(\tau_k(u) \wedge t)$ because $M_k$ is continuous). However, the content of the assertion lies in the fact that the $B_k(u)$ remain, in some sense, independent Brownian motions. Here it is necessary to allow for the fact that $B_k(u)$ is only defined for $u \le \sup_t \langle M_k \rangle_t$, which may be finite. We can again extend by a product-space construction, as in the proof of Theorem 1.1, to obtain independent Brownian motions defined for all t. However, it is necessary to point out that the processes $B_k(u \wedge \langle M_k \rangle_t)$ are not stopped Brownian motions with respect to the filtrations of these extended processes, in the sense that

$\langle M_k \rangle_t$ need not be optional with respect to these $B_k$. We can only assert that it is optional with respect to the usual time-changed filtration $\mathcal{Z}_{\tau_k(t)}$, which follows by a routine verification. Furthermore, the integrands $h_k(\tau_k(u-))$ are previsible only with respect to these filtrations [16, loc. sit., Lemma II.1], which in general depend on $k$. We will discuss this in more detail below, since it is a special case of treating H3.

It will be convenient to have a special terminology for this kind of time-limited process, as follows.

DEFINITION 2.2. Let $(Y_k(t), k < N+1)$, $N \leq \infty$, be processes defined on the same space $(\Omega, \mathcal{Z}, P)$. We say that $(Y_k)$ is a *halted* N-dimensional Lévy process if

a) there are measurable $0 \leq T_k \leq \infty$ such that $Y_k(t) = Y_k(t \wedge T_k)$, $0 \leq t$, and

b) there is a sequence $(W_k;\ W_k(0) = 0,\ k < N+1)$ of independent Lévy processes (processes with homogeneous independent increments) defined on a separate space such that, if we construct the product probability space $(\Omega^*, \mathcal{Z}^*, P^*)$ and on it define $Y_k^*(t) = Y_k(t \wedge T_k) + W_k(t - (t \wedge T_k))$, $t \geq 0$, then $(Y_k^*, k < N+1)$ is a sequence of independent Lévy processes on the product space, with $Y_k^*(0) = 0$.

In order for this definition to be useful, one needs to have

THEOREM 2.3. *Except in the trivial case* $P\{T_k = 0\} = 1$, *the law of* $Y_k^*$ *is uniquely determined by that of* $(Y_k, T_k)$, *and (unless* $P\{T_k = \infty\} = 1$) *it coincides with the law of* $W_k$.

PROOF. Let $\ln E(\exp iuY_k^*(1)) = \psi(u) = i\gamma u - \delta \frac{u^2}{2} + \int \left(e^{iux} - 1 - \frac{iux}{1+x^2}\right) \frac{1+x^2}{x^2}\, dG(x)$ be the Lévy canonical form for $Y_k^*$, where $G(0+) - G(0-) = 0$, $\delta \geq 0$, $G(\infty) - G(-\infty) < \infty$ (see for example

[6, Chap. 3, §18]). It is well known that the parameters $\lambda$, $\delta$, and dG are determined uniquely by a generic path of $Y^*$ in $t > K$ for any K. For example, with probability 1, $G(\infty) - G(x) = \lim_{K\to\infty} \lim_{t\to\infty}(t^{-1}(\#\{t_j \in (K;K+t): Y_k^*(t_j) - Y_k^*(t_j-) \geq x\})$. From this we see immediately that, unless $P\{T_k = \infty\} = 1$, the processes $Y_k^*$ and $W_k$ have the same law. It then follows that the law of $Y_k^*$ is unchanged if we *condition* on $\{T_k > 0\}$, since this is true given $\{T_k = 0\}$. To complete the proof, it suffices to observe that $\lim_{\varepsilon\to 0} \varepsilon^{-1}(\varepsilon - (\varepsilon \wedge T_k)) = 0$ on $\{T_k > 0\}$, so that

$$
\begin{aligned}
(2.2)\qquad \psi(u) &= \lim_{\varepsilon\to 0} \varepsilon^{-1} \ln E(E(\exp(iu\, Y_k^*(\varepsilon))|T_k);\ T_k > 0)P^{-1}\{T_k > 0\} \\
&= \lim_{\varepsilon\to 0} \varepsilon^{-1} \ln E(E(\exp(iuY_k(\varepsilon)\exp(\psi(u)(\varepsilon - \varepsilon\wedge T_k))|T_k); \\
&\qquad T_k > 0)P^{-1}(T_k > 0) \\
&= \lim_{\varepsilon\to 0} \varepsilon^{-1} \ln E(\exp(iuY_k(\varepsilon))(1 + 0(\varepsilon - \varepsilon\wedge T_k));\ T_k > 0)P^{-1}(T_k > 0) \\
&= \lim_{\varepsilon\to 0} \varepsilon^{-1} E(\exp\, iuY_k(\varepsilon);\ T_k > 0)P^{-1}(T_k > 0)\cdot
\end{aligned}
$$

where we used the boundedness, for each u, of $\varepsilon^{-1}|1 - \exp(\psi(u)(\varepsilon - \varepsilon\wedge T_k))|$ as $\varepsilon \to 0$.

REMARK. In most applications below we have $T_k = \inf\{t: Y_k(t+s) = Y_k(t)$ for $s > 0\}$. Then the law of $Y_k^*$ is uniquely determined by that of $Y_k$. However, simple examples show that this is not always true when $T_k$ is unspecified.

We turn now to extending (2.1) under H3. Following the notation of [15], we denote by $\mathfrak{M}_s^2$ the closed subspace of $\mathfrak{M}_0^2$ consisting of *strict* martingales, where $M_s(t)$ is strict if and only if, for all optional T, we have $M_s(T) \in \mathcal{Z}_{T-}$. It is shown in [15] that the orthogonal complement of $\mathfrak{M}_s^2$ is the space $\mathfrak{M}_j^2$ of "pure jump" martingales, which is the closure in $\mathfrak{M}_0^2$ of the space generated by the single jump martingales

$\phi I_{\{T\le t\}}$; $E(\phi|\mathcal{Z}_{T-}) = 0$, $E\phi^2 < \infty$. Consequently, H3 is equivalent to $\mathcal{M}_0^2 = \mathcal{M}_s^2$. As usual, let $\mathcal{M}_c^2$ denote the subspace of continuous martingales, and its orthogonal complement $\mathcal{M}_d^2$ the subspace of "purely discontinuous" ones. Then under H3, we have $\mathcal{M}_s^2 = \mathcal{M}_c^2 \oplus \mathcal{M}_d^2$ (of course, analogous statements hold also for $\mathcal{M}_0^2(t)$). By the familiar orthogonalization procedure of Kunita-Matanabe [13], we may choose a basis $M_1^c,\ldots,M_{n_c}^c$ of $\mathcal{M}_c^2$, as well as a basis $M_1^d,\ldots,M_{n_d}^d$ of $\mathcal{M}_d^2$. It is also known (see Davis and Varaiya [3], Th. 1]) that we may impose the further condition

$$\text{(2.3)}\qquad d\langle M_1^c\rangle_t >> d\langle M_2^c\rangle_t >>\cdots, \quad\text{and}\quad d\langle M_1^d\rangle_t >> d\langle M_2^d\rangle_t >>\cdots,$$

in which case the numbers $n_c(= n_c(\infty))$ and $n_d$ are minimal, and called the 2-dimensions of $\mathcal{M}_c^2$ and $\mathcal{M}_d^2$ (and also $n_c(t)$, $n_d(t)$, etc. for $\mathcal{M}_0^2(t)$). We do not require this in the case of $\mathcal{M}_d^2$, however.

It should be noted that under H3 the 2-dimension of $\mathcal{M}_0^2$ is *not*, in general, $n_c + n_d$. On the other hand, our extension of (2.1) under H3 below suggests that the 2-dimension of $\mathcal{M}_0^2$ is also not the best strict-sense extension of $E(\infty)$, and in general *neither* is $n_c + n_d$. We can state our main theorem as follows (see also Corollary 2.6, and for a converse Theorem 2.9 below).

THEOREM 2.4. *Under* H3, *representation* (2.1) *remains valid if* a) *we replace* $N(t)$ *by* $n_c(t) + n_p(t)$, *where* $n_p(t)$ $(\ge n_d(t))$ *is a suitably defined Poisson index, and* b) *we permit compensated Poisson martingales* $P_k(u)$ *as well as Brownian motions* $B_k(u)$. *More precisely, we have*

$$\text{(2.4)}\qquad X = \sum_{i<n_c(t)+1} \int_0^\infty h_i(\tau_i^c(u-))dB_i(u\wedge\langle M_i^c\rangle_t) + \sum_{j<n_p(t)+1} \int_0^\infty k_j(\tau_j^d(u-))dP_j(u\wedge\langle M_j^d\rangle_t)$$

*where*

$$(B_i(u \wedge \langle M_i^c \rangle_t),\ i < n_c(t)+1; P_j(u \wedge \langle M_j^d \rangle_t),\ j < n_p(t)+1))$$

*for each* $t$ *is a halted* $n_c(t)+n_p(t)$*-dimensional Lévy process, such that* $(B_i(u \wedge \langle M_i^c \rangle_t;\ i < n_c(t)+1)$ *halted Brownian motion,* $P_j(u \wedge \langle M_j^d \rangle_t$, $j < n_p(\infty)+1)$ *is a halted compensated Poisson process* (i.e. *each process* $P_j^*(u)+u$ *as in Definition 2.2, is a standard Poisson process with* $\lambda = 1$). *The* $\tau_i^c$ *and* $\tau_j^d$ *are right-continuous inverses of* $\langle M_i^c \rangle$ *and* $\langle M_j^d \rangle$ *as in Theorem* 2.1, *where the set* $\{M_i^c, M_j^d\}$ *forms an orthogonal basis for* $\mathfrak{M}_0^2$, $B_i(u \wedge \langle M_i^c \rangle_t) = M_i^c(\tau_i^c(u) \wedge t)$, $P_j(u \wedge \langle M_j^d \rangle_t) = M_j^d(\tau_j^d(u) \wedge t)$. *The integrands* $h_i$ *and* $k_j$ *are again* $\mathcal{Z}_t$*-previsible* (*only these depend on the choice of* $X \in L_0^2(\mathcal{Z}_t)$).

PROOF. We first of all transfer the entire problem to a canonical space, as was done in the proof of Lemma 1.5. Recalling that $L_0^2(\Omega, \mathcal{Z}_\infty, P)$ is separable (assumption b') of Section 1), let $(X_n,\ 1 \le n)$ be any orthonormal basis and consider the martingales $M_n(t) = E(X_n | \mathcal{Z}_t)$, chosen to be right-continuous with left limits. It is an elementary exercise to show that the filtration generated by $(M_n(s),\ n \ge 1,\ s \le t)$ differs from $\mathcal{Z}_t$ only by P-null sets for each $t$. Indeed, for any $X \in L_0^2(\Omega, \mathcal{Z}_\infty, P)$ there are linear combinations $Y_n = \sum c_{i,n} X_n$ converging in $L^2$ to $X$, hence by a classical martingale maximal inequality of Doob there is a subsequence $n_k \to \infty$ such that $E(Y_{n_k} | \mathcal{Z}_t) - E(X | \mathcal{Z}_t)$ converges to 0 as $n_k \to \infty$ uniformly in $t$ with probability one. Thus $E(X | \mathcal{Z}_t)$ is in the generated $\sigma$-field up to P-null sets. A little reflection now reveals that it suffices to prove Theorem 2.4 on the canonical space of all sequences $(w_n(t),\ 1 \le n)$ of right-continuous paths with left limits, with $w_n(0) = 0$ for all $n$, and with generated $\sigma$-fields given the joint law of $(M_n(t))$, augmented and completed in the usual way. Indeed, the mapping $w \to (M_n(t,w))$ has range of outer measure 1 in the canonical space, and its measure-preserving inverse suffices to transfer the representation (2.4) back to the original probability space. It is also clear that H3 transfers to the canonical space (as usual, we treat indistinguishable processes, and in particular those indistinguishable from 0, as identical). Consequently, in proving the theorem we henceforth assume the canonical situation, and without change of notation continue to use $X$, $\mathcal{Z}_t$, $P$, etc. for their canonical images.

We now proceed as in the proof of Lemma 1.5 to introduce the prediction process $Z_t(S)$; $S \in \mathcal{Z}^0$ of $P$ (this was denoted $Z_t^P$ in [11, Essay 1, Definition 1.6]). As before, we consider it on the compactification $H_0^+$ of the non-branching points, with Borel field $\mathcal{H}$ and the Ray topology, so that the times of discontinuity of $Z_t$ are the same (up to a fixed P-null set) as those of the martingales

$$(2.5) \qquad M_{f,\lambda}(t) = R_\lambda^Z f(Z_t) - R_\lambda^Z f(Z_0) + \int_0^t f(Z_s) - \lambda R_\lambda^Z(f(Z_s))ds$$

when $f$ varies in a countable uniformly dense set of continuous functions on $H_0^+$. On the other hand, by H3 these martingales are in $\mathcal{Z}_s^2$ up to finite $t$, and hence (by [16, Prop. 1.2c)]) if $T$ is any (finite) previsible stopping time then $Z_{T-} = Z_T$, P-a.s., where $Z_{t-}$ denotes the "left-limit process" of (1.3) and we appeal to the section theorem to show that the left limits of $R_\lambda^Z f(Z_t)$ are given by $R_\lambda^Z f(Z_{t-})$. Thus the discontinuity times of $Z_t$ are totally inaccessible. It also follows by the result of Kunita-Watanabe (according to which the martingales (2.5) generate $\mathfrak{M}_0^2(t)$ for finite $t$) that *any* $M(t) \in \mathfrak{M}_0^2$ has its times of discontinuity contained in those of $Z_t$, P-a.s. Finally, the process $Z_t$ generates the same augmented $\sigma$-fields $\mathcal{Z}_t$ as does $\mathcal{Z}_{t+}^0$ ([11, Essay I, Theorem 1.2]), hence we can reduce the study of purely discontinuous martingales of $(\mathcal{Z}_t, P)$ to a study of $Z_t$ (note for example that under

H3, since $M(T) \in \mathcal{Z}_{T-}$ for any $M \in \mathfrak{M}_0^2$ and optional $T$, it follows from expression (2.5) that $Z_T \in \mathcal{Z}_{T-}$ as well).

Our main reason for introducing $Z_t$ into the proof, apart from what seems to be its inevitable relationship, is to make use of the theory of Lévy systems of a Markov process to treat the discontinuous martingales of $\mathcal{Z}_t$.

LEMMA 2.5. *Under* H3, *the space* $\mathfrak{M}_d^2$ *is generated (in the sense of stochastic integrals) by a single martingale of the form*

$$\hat{M}_d(t) = \sum_{s \le t} f(Z_{s-}, Z_s) - \int_0^t dH(s) \int_{H_0} N_Z(Z_{s-}, dz) f(Z_{s-}, z)$$

*with* $0 \le f(z_1, z_2) \in \mathcal{H} \times \mathcal{H}$, $f(z,z) = 0$, *where* $(H(s), N_Z(z_1, dz))$ *is a Lévy system for* P *in the sense of* [11, Essay IV, Definition 2.2].

REMARKS. It is not asserted that $\hat{M}_d \in \mathfrak{M}_d^2$, but only that $\hat{M}_d \in \mathfrak{M}_d^2(t)$ for every finite $t$. It follows, however, that the 2-dimension of $\mathfrak{M}_d^2$ is at most 1 (since $\int_0^\infty h(t) d\hat{M}_d(t) \in \mathfrak{M}_d^2$ for a suitable non-random $h(t) > 0$). Note further than $Z_{s-}$ refers to the left-limit process of (1.3), and that $H(s)$ is continuous. Since the times of discontinuity of $Z_s$ are totally inaccessible, we know (for example, [11, Essay I, Theorem 2.13 (ii)]) that, with probability 1, $Z_{T-} \in H_0$ (= non-branch points) at all times of discontinuity $T$. Accordingly the "discontinuous part" $(L(s), M_Z)$ of the Lévy system plays no role under H3. The reader who refers to [11] for $(H, N_Z)$ should note that in (2.1)(a), p. 93, $Z_s^h$ should read $Z_{s-}^h$ at one point. In any case, this is just a transcription of the Lévy system of the Ray process $Z_t$ to a different probability space and to the state space H of probabilities (rather than the Ray space $H_0^+$; for details, see [11, Essay IV, Theorem 1.2]).

PROOF: Let $0 \le f_n \le 1$ be a sequence uniformly dense in the band of $C(H_0^+)$ satisfying these inequalities. Then the times of discontinuity of $Z_t$ are contained in those of the set $\{M_{f_n,\lambda}(t);\ 1 \le n,\ \lambda$ rational$\}$. It follows by decomposing into continuous and purely discontinuous parts [4, VIII, [43]] and using (for example) identity of [12, Theorem 3.1] that

$$\begin{aligned} EM^2_{f,\lambda}(t) &= E[(R^Z_\lambda f(Z_t))^2 - (R^Z_\lambda f(Z_0))^2 \\ &\qquad + 2\int_0^t (f(Z_s) - \lambda R^Z_\lambda f(Z_s))R^Z_\lambda(f(Z_s))ds] \\ &\le \lambda^{-2} + 4t\lambda^{-1} \le \left(\frac{\lambda+1}{\lambda}\right)^2 (1+2t) \end{aligned}$$

if $\sup|f| \le 1$, so that we have for each $(\lambda,n)$,

$$\left(\frac{\lambda+1}{\lambda}\right)^2 (1+2t) \ge E\sum_{s\le t} (R^Z_\lambda f_n(Z_{s-}) - R^Z_\lambda f_n(Z_s))^2. \tag{2.6}$$

Then letting $(\lambda,n)_i$, $1 \le i$, be an enumeration of the pairs $((\lambda,n);$ $0<\lambda$ rational), and moreover letting $\lambda_j$ be a separate enumeration of rationals $\lambda > 0$, we see that if $(\lambda(z_1,z_2),\ n(z_1,z_2))$ is the pair $(\lambda,n)_i$ for the smallest $i = i(z_1,z_2)$ such that $R^Z_\lambda f_n(z_1) \ne R^Z_\lambda f_n(z_2)$ at $(\lambda,n) = (\lambda,n)_i$, and if $j(\lambda_j) = j$, then

$$\begin{aligned} f(z_1,z_2) = 2^{-(n(z_1,z_2)+j(\lambda(z_1,z_2)))} \frac{\lambda(z_1,z_2)}{1+\lambda(z_1,z_2)} &|R^Z_{\lambda(z_1,z_2)} f_{n(z_1,z_2)}(z_1) \\ &- R^Z_{\lambda(z_1,z_2)} f_{n(z_1,z_2)}(z_2)| \end{aligned}$$

satisfies $0 < f < 1$ and

$$\begin{aligned} E\sum_{s\le t} f^2(Z_{s-},Z_s) &\le E\sum_{s\le t}\sum_{j,n} 2^{-(n+j)}\left(\frac{\lambda_j}{1+\lambda_j}\right)^2 \\ &\qquad \times (R^Z_{\lambda_j} f_n(Z_{s-}) - R^Z_{\lambda_j} f_n(Z_s))^2 \\ &\le (1+2t). \end{aligned}$$

It is then clear that the martingale $\hat{M}_d$ defined from this $f(z_1,z_2)$ as in Lemma 2.5 is square integrable, and its times of discontinuity are the same as those of $Z_t$ (outside a P-null set). Finally, to see that every element of $\mathfrak{M}_d^2$ has the form $\int_0^t h(s)d\hat{M}_d(s)$, $h(s)$ previsible, we note that if this were false there would be an element $M_d \in \mathfrak{M}_d^2$ orthogonal to $\hat{M}_d$, in the sense of stochastic integration. This is impossible because such an element $M_d$ would have its times of discontinuity contained in those of $\hat{M}_d$. Hence $[M_d,\hat{M}_d]_t = \sum_{0\le s\le t} (\Delta M_d(s)\Delta\hat{M}_d(s))$ could not be indistinguishable from 0 [4, VII, [37]], while on the other hand it would necessarily be a martingale (since $\langle M_d,\hat{M}_d\rangle_t = 0$). But, by definition, $|\Delta\hat{M}_d(s)| < 1$ holds for all $s$, hence

$$E \sum_{0\le s\le t} (\Delta M_d(s)\Delta\hat{M}_d(s))^2 \le E \sum_{0\le s\le t} (\Delta M_d(s))^2 = EM_d^2(t) < \infty.$$

Thus we would have, for suitable non-random $h(t) > 0$, $\int_0^\infty h(t)d[M_d,M_d]_t \in \mathfrak{M}_j^2$, contrary to H3.

REMARK. The construction of $M_d \in \mathfrak{M}_d^2(t)$ having the same times of discontinuity as $Z_t$ made no use of H3, or the assumed degeneracy of $\mathcal{Z}_0$. It is essentially a well-known consequence of the Lévy system, and it is valid for any P on the canonical space, as is also the observation that the discontinuity times of $Z_t$ are equivalent to those of any countable family of martingales generating $\mathfrak{M}_0^2$. We needed H3 only to the effect that the two martingales cannot be orthogonal and have times of discontinuity in common. It is plausible that under H3 this is true of *any* two elements of $\mathfrak{M}_0^2$, but we do not know a proof and the result is not needed here.

Returning to the proof of Theorem 2.4 we will define the martingales $M_j^d(t)$ from $\hat{M}_d$, and at the same time obtain the *minimal* Poisson index $n_p(t)$. It emerges that $n_p(t) = 0,1$, or $\infty$. In the first place, if $\mathfrak{M}_0^2(t) = \mathfrak{M}_c^2(t)$ then $n_p(t) = 0$ and the result (up to time t) was already proved. Next fix t and set $g(z_1,z_2) = I_{\{z_1 \neq z_2\}}$.

*Case* 1. $E \sum_{s \leq t} g(Z_{s-},Z_s) < \infty$.

In this case, let $f_n(z_1,z_2) = I_{\{f(z_1,z_2) \leq 1/n\}}$. It follows that for each n

$$\hat{M}_n(u) = \sum_{s \leq u} f_n(Z_{s-},Z_s) - \int_0^u dH(s) \int_{H_0} N_Z(Z_{s-},dz) f_n(Z_{s-},z)$$

is in $\mathfrak{M}_d^2(t)$ with $E\langle \hat{M}_n \rangle_t = E \sum_{s \leq t} f_n(Z_{s-},Z_s) \leq E \sum_{s \leq t} g(Z_{s-},Z_s)$. Letting $n \to \infty$ we obtain $M_1^d(u) = \sum_{s \leq u} g(Z_{s-},Z_s) - \int_0^u dH(s) \int_{H_0} N_Z(Z_{s-},dz) g(Z_{s-},z)$ (def.) in $\mathfrak{M}_d^2(t)$, with the same times of discontinuity as $\hat{M}_d$, and all jumps of unit size. To show that $M_1^d(u)$ generates $\mathfrak{M}_d^2(t)$ it suffices to exhibit a representation $\hat{M}_d(u) = \int_0^u h(v) dM_1^d(v)$, $0 < u \leq t$. Here we set

$$h(v) = \frac{\int_{H_0} N_Z(Z_{v-},dz) f(Z_{v-},z)}{\int_{H_0} N_Z(Z_{v-},dz) g(Z_{v-},z)}, \tag{2.7}$$

(or $h(v) = 0$ if undefined). It follows by an immediate calculation that the continuous part (compensator) of $\int_0^u h(v) dM_1^d(v)$ is the same as that of $\hat{M}_d(u)$. Moreover, by Schwartz' inequality we have

$$\begin{aligned} E\left\langle \int_0^u h(v) dM_1^d(v) \right\rangle_t &= E \int_0^t h^2(v) d\langle M_1^d \rangle_v \\ &\leq E \int_0^t \left( \frac{\int_{H_0} N_Z(Z_{v-},dz) f^2(Z_{v-},z)}{\int_{H_0} N_Z(Z_{v-},dz) g(Z_{v-},z)} \right) d[M_1^d]_v \\ &= E \int_0^t dH(v) \left( \int_{H_0} N_Z(Z_{v-},dz) f^2(Z_{v-},z) \right) = E\langle \hat{M} \rangle_t \end{aligned}$$

so it follows that $\int_0^u h(v) dM_1^d(v) \in \mathfrak{M}_0^2(t)$. But then $\hat{M}_d(u) - \int_0^u h(v) dM_1^d(v) \in \mathfrak{M}_j^2$ [16, Prop. III. 4], so according to H3 this must be 0, as asserted.

Now setting $n_p(t) = 1$, we may represent any $X \in L_0^2(\mathcal{Z}_t)$ in the form

$$(2.8) \qquad X = \sum_{i<n_c(t)+1} \int_0^t h_i(s)dM_i^c(s) + \int_0^t k_1(s)dM_1^d(s),$$

for previsible $h_i$ and $k_1$. Since $M_1^d$ has jumps, we must use a bit of care in making the time change using $\tau_1^d(u) = \inf\{s: \langle M_1^d\rangle_{s\wedge t} > u\} (\le \infty)$. Then Proposition 3 of Le Jan [16] used before requires that we *define* right-continuous $P_1(u) = M_1^d(\tau_1^d(u) \wedge t)$. Except for this and an obvious application of the monotone class theorem to extend [16] to unbounded integrands $h_i$ and $k_1$, the formal change of variables is immediate. Observing that, for $u > 0$, $\int_0^\infty k_1(\tau_1^d(u-))dP_1(u\wedge \langle M_1^d\rangle_t) = \int_0^\infty k_1(\tau_1^d(u-))dM_1^d(\tau_1^d(u)\wedge t)$ we obtain

$$(2.9) \qquad X = \sum_{i<n_c(t)+1} \int_0^\infty h_i(\tau_i^c(u-))dB_i(u\wedge\langle M_i^c\rangle_t) + \int_0^\infty k_1(\tau_1^d(u-))dP_1(u\wedge\langle M_1^d\rangle_t).$$

We need to check that $(B_i(u\wedge\langle M_i^c\rangle_t), P_1(u\wedge\langle M_1^d\rangle_t))$ is a halted $(n_c(t)+1)$-dimensional Lévy process, with Brownian laws for the $B_i$, and Poisson law for $P_1$. Neither of these assertions is really new. If we attach a standard Poisson martingale $W_1^d$ via a product space, as in the proof of Theorem 1.1, we obtain a martingale

$$M_1^{d*}(s) = \begin{cases} M_1^d(s) & \text{if } s \le t \\ M_1^d(t) + W_1^d(s-t) & \text{if } t < s, \end{cases}$$

which is obviously purely discontinuous and has only unit jumps. Further, its times of discontinuity are obviously totally inaccessible in the product space filtration

$$\mathcal{Z}_s^* = \begin{cases} \mathcal{Z}_s\ ; & s \le t \\ \mathcal{Z}_t \vee \mathcal{Z}^d_{s-t}\ ; & t < s\ . \end{cases}$$

The same remains true if we simultaneously attach Wiener processes $W_i^c$, $i < n_c(t) + 1$, and define $M_i^{c*}(s)$ analogously. Then these added martingales are continuous. It is well known that the processes

$$M_i^{c*}(\tau_i^{c*}(u));\ \tau_i^{c*}(u) = \inf\{v : \langle M_i^{c*}\rangle_v > u\}$$

are Brownian motions, and also (by a result of S. Watanabe [19]) that

$$M_1^{d*}(\tau_1^{d*}(u));\ \tau_1^{d*}(u) = \inf\{v : \langle M_1^{d*}\rangle_v > u\}$$

is a compensated Poisson process with $\lambda = 1$. We have remarked previously that the former are independent. Finally, in [18] P. A. Meyer gave another proof of this fact, and moreover extended it to the case of an orthogonal n-tuple of purely discontinuous martingales with totally inaccessible jumps all of size 1, thus obtaining by time changes an n-tuple of compensated Poisson processes ([18, Theorem 2', p. 195]). But one need only read the last two paragraphs of [18] to realize that his proof applies without change to the "mixed" case of (finitely many) continuous martingales and (finitely many) purely discontinuous ones. Thus we can finally respond to the question at the end of [18] to the effect that, if Theorem 2' of [18] is merely a "curiosité mathématique," then so is the present Theorem 2.4 since the former provides a key step of its proof. To complete the case $n_p(t) = 1$, we now remark that, because

$$\tau_1^{d*}(u) = \tau_1^d(u \wedge \langle M_1^d\rangle_t) + (u - (u \wedge \langle M_1^d\rangle_t)),$$

we have also

$$M_1^{d*}(\tau_1^{d*}(u)) = M_1^d(\tau_1^d(u \wedge \langle M_1^d\rangle_t)) + W_1^d(u - (u \wedge \langle M_1^d\rangle_t)),$$

and analogously for $M_i^{c*}$, showing that

$$(B_i(u \wedge \langle M_i^c \rangle_t),\ i < n^c(t)+1;\ P_1(u \wedge \langle M_1^d \rangle_t))$$

is a halted Lévy process (Definition 2.2).

*Case* 2. $E \sum_{s \le t} g(Z_{s-}, Z_s) = \infty$.

In this case it is clear that we must have $n_p(t) = \infty$. Indeed, each $\langle M_j^d \rangle_s$ is a stopping time of the corresponding $\mathcal{Z}_{\tau_j^d(s)}$, whence by the optional stopping theorem for $P_j$ we have $E P_j^2(\langle M_j^d \rangle_t) = E\langle M_j^d \rangle_t < \infty$. The left side is the expected number of jumps of $P_j(u \wedge \langle M_j^d \rangle_t)$, and all the jumps of $\hat{M}_d$, $s \le t$, are contained in these (since $\hat{M}_d(s) = E(\hat{M}_d(t) | \mathcal{Z}_s)$ is to be represented as in (2.4) with s in place of t). If $n_p(t)$ were finite, this would contradict our assumption. However, let us set

$$g_1(z_1, z_2) = I_{\{f(z_1,z_2) \le 1\}}, \quad \text{and for } j > 1,$$

$$g_j(z_1, z_2) = I_{\{j^{-1} \le f(z_1,z_2) < (j-1)^{-1}\}}.$$

Then the martingales

$$M_j^d(u) = \sum_{s \le u} g_j(Z_{s-}, Z_s) - \int_0^u dH(s) \int_{H_0} N_Z(Z_{s-}, dz) g_j(Z_{s-}, z))$$

are in $\mathfrak{M}_d^2(t)$ and have no times of discontinuity in common. Hence they are orthogonal [4, VIII, [43]d)]. It follows in exactly the same way as for Case 1 that $(B_i,\ i < n_c(t)+1;\ P_j,\ j < \infty)$ in (2.4) is a halted Lévy process with Brownian and compensated Poisson components, so it remains only to check the representation of arbitrary $X \in L_0^2(\mathcal{Z}_t)$. As in Case 1, this reduces to exhibiting a representation $\hat{M}_d(u) =$

$\sum_j \int_0^u h_j(v)dM_j^d(v)$, since then we have only to change variables in each term of (1.2) as before. Moreover, we need only substitute $g_j$ for g and $fg_j$ for f in (2.7) to define $h_j$ such that $\int_0^u h_j(v)dM_j^d(v) = \int_0^u g_j(v)d\hat{M}_d(v)$, as follows again by noting that both sides are in $\mathfrak{M}_0^2(t)$ and they have equal continuous parts. Thus the proof is concluded, and we have in addition

COROLLARY 2.6. *The minimal value of* $n_p(t)$ *in* (2.4) *is* 0,1, *or* $\infty$. *We may use the same halted Lévy process* $(B_i(u \wedge \sup_s \langle M_i^c \rangle_s)$, $P_j(u \wedge \sup_s \langle M_j^d \rangle_s))$ *for all* t, *provided that we set* $B_i(u) = 0$ *for* $i > n_c(u)$ *and* $P_j(u) = 0$ *for* $j > n_p(u)$ *(minimal), and permit* $\langle M_1^d \rangle_t = \infty$.

Indeed, if $t_0 = \sup\{t: n_p(t) \le 1\}$, then, because $\mathcal{Z}_t$ is quasi-left-continuous under H3, we can write $E(X|\mathcal{Z}_{t_0})$ as in (2.4) with $n_p(t_0) = 1$ by allowing $\langle M_1^d \rangle_{t_0} = \infty$, and then on $\{\langle M_1^d \rangle_{t_0} < \infty\}$ *redefine* $P_1(u)$ for $u > \langle M_1^d \rangle_{t_0}$ using $P_1(\langle M_1^d \rangle_{t_0})$ + ($\Delta P_1$ from Case 2). Thus, we use Case 2 only to represent $E(X|\mathcal{Z}_t) - E(X|\mathcal{Z}_{t_0})$ for $t_0 < t$, which is consistent with the redefined $P_1$ since it involves only $dP_j(u)$ for $u > t_0$. It can be seen that our definition makes

$$(P_j((u + \langle M_j^d \rangle_{t_0}) \wedge \langle M_j^d \rangle_t) - P_j(\langle M_j^d \rangle_{t_0}),\ 1 \le j)$$

a halted Poisson process for each $t > t_0$, and its continuation is independent of $\mathcal{Z}_{t_0}$ (i.e. it is still compensated Poisson given $\mathcal{Z}_{t_0}$) in such a way that, finally,

$$(B_i(u \wedge \sup_s \langle M_i^c \rangle_s),\ i < n_c(\infty) + 1;\ P_j(u \wedge \sup_s \langle M_j^d \rangle_s,\ j < n_d(\infty) + 1)$$

is also a halted Lévy process.

We will point out two more results complementary to Theorem 2.4. For the first, it is useful to introduce

DEFINITION 2.7. We say that $L_0^2(\mathcal{Z}_t)$ has a *halted Lévy process representation* if (2.4) holds when $(B_i,P_j)$ is replaced by an arbitrary halted Lévy process, with $n_c(t) + n_p(t)$ replaced by arbitrary $n(t)$, and orthogonal martingales $(M_k(s), k < n(t) + 1)$ as before.

PROPOSITION 2.8. *Under* H3, *every halted Lévy process representation involves only Brownian, compensated Poisson, or Brownian-plus-compensated Poisson terms. Moreover,* $n_c(t) + n_p(t)$ *cannot be reduced by any such representation.*

PROOF. For fixed t, the family $\mathcal{Z}_{s\wedge t}$ is right-continuous and, setting $M = M_k$, $\langle M\rangle_{s\wedge t}$ is adapted. According to [16, Prop. 3], if $N_s \in \mathfrak{M}_0^2(\mathcal{Z}_{t\wedge s})$, then $N_{\tau(s)} \in \mathfrak{M}_0^2(\mathcal{Z}_{\tau(s)\wedge t})$ where $\tau(s)$ is the right-continuous inverse $(\le \infty)$ of $\langle M\rangle_{s\wedge t}$. Conversely, if we start with any $N_s \in \mathfrak{M}_0^2(\mathcal{Z}_{\tau(s)\wedge t})$, then $N_{\langle M\rangle_{s\wedge t}} \in \mathfrak{M}_0^2(\mathcal{Z}_{s\wedge t})$, [16,(10)]. It is clear that these mappings preserve both orthogonality and the property of being a pure-jump martingale. Therefore, under H3 it follows that $M_{\tau(s)}$ is strict, for otherwise the $M_{\tau\langle M\rangle_{s\wedge t}}$ would not be strict either. But under the conditions of the proposition, $M_{\tau(s)}$ would then be a strict halted Lévy process. It follows easily that the Lévy-measure of $M^*_{\tau(s)}$ (Definition 2.2) must either be void or a single point measure. Otherwise, the time $T_\varepsilon$ of the first jump exceeding $\varepsilon > 0$ would not satisfy $M^*_{\tau(T(\varepsilon))} \in \mathcal{Z}^*_{\tau(T_\varepsilon-)}$ for small $\varepsilon$ (see also He and Wang [8, Theorem 2.2]).

By permitting such Lévy processes, it is clear that if $n_p(t) = \infty$ (i.e. $E(\#\text{ jumps before } t) = \infty$) then there must always be infinitely many terms in (2.4) (since necessarily $E(\#\text{ jumps of } M_k \text{ before } t) < \infty$). Thus we may assume $n_p(t) = 1$, and also $n_c(t) < \infty$ since clearly to represent $\mathfrak{M}_c^2$ we need at least $n_c(t)$ orthogonal summands (or equivalently, their continuous components, which need not be orthogonal).

Now if the same $n_c(t)$ elements of $\mathfrak{M}_0^2(t)$ suffice also to represent $M_1^d$, then we obtain as by-product a representation of 0 among their continuous components, showing that in any case at least one more term is required (the continuous components would only generate a subspace of 2-dimension at most $n_c(t)-1$). Thus we require at least $n_c(t)+1$ terms, as was to be shown.

We conclude by proving a converse to Theorem 2.4 which may make its meaning a little clearer.

THEOREM 2.9. *Let* $\mathcal{Z}_t$ *be a filtration defined under conditions* a') *and* b') *of Section* 1, *and suppose that the conclusion of Theorem* 2.4 *holds, using a fixed halted Brownian and compensated Poisson process as in Corollary* 2.6 *(but without assuming minimal values for* $n_c(t)$ *and* $n_p(t)$). *Then hypothesis* H3 *holds.*

PROOF. In this case we cannot assume a priori that $\langle M_j^d\rangle_t$ are continuous. However, $\tau_j(u)$ are stopping times of $\mathcal{Z}_t$, in such a way that $\int_0^s k_j(\tau_j(u))dP_j(u\wedge\langle M_j^d\rangle_t)$ is a square integrable martingale with respect to $\mathcal{Z}_{\tau_j(s)}$ (and analogously for $B_i(u\wedge\langle M_i^c\rangle_t)$). If H3 did not hold, there would be a single jump martingale of $\mathcal{Z}_t$, and we may suppose the jump precedes t with positive probability. Let X denote this martingale at time t, so that for $s<t$ it is given by $E(X|\mathcal{Z}_s)$. If we write the expression (2.4) for X, then even if the series is infinite we can write $E(X|\mathcal{Z}_s)$ term-by-term, as the series converges in $L_0^2(\mathcal{Z}_t)$ by Jensen's inequality. On the other hand, for each j two more applications of the inverse change of variables [16, (10)] yield for $s<t$

$$(2.10)\qquad E\Big(\int_0^\infty k_j(\tau_j^d(u-))dP_j(u\wedge\langle M_j^d\rangle_t)\,|\mathcal{Z}_s\Big)$$

$$=\int_0^{\langle M_j^d\rangle_s} k_j(\tau_j^d(u-))dP_j(u\wedge\langle M_j^d\rangle_t)$$

$$= \int_0^\infty k_j(v)dM_j^d(\tau_j(\langle M_j^d\rangle_{(v\wedge s)})) = \int_0^s k_j(v)dM_j^d(v)$$

(this does not require continuity of $\langle M_j^d\rangle_v$). It follows that these martingales (and also their analogs with $B_i(u \wedge \langle M_i^c\rangle_t)$) are orthogonal in $\mathfrak{M}_0^2(t)$. Moreover, since the jumps of $M_j^d$ all have size 1, if $T$ is a time of discontinuity of (2.10) then the jump at $T$ is of size $k_j(T)$, where $k_j(v)$ is $\mathcal{Z}_v$-previsible. Then $k_j(T) \in \mathcal{Z}_{T-}$ (by [4, Vol. I, IV, 67]) from which it follows that the above martingale terms are in $\mathfrak{M}_s^2(t)$. On the other hand, their sum reduces to a single jump, hence it is in $\mathfrak{M}_j^2(t)$. $\mathfrak{M}_s^2(t)$ being closed, we have a contradiction which completes the proof.

## References


1. H. CRAMER. On some classes of non-stationary stochastic processes. *Proc. of the Fourth Berkeley Symposium II,* J. Neyman, Ed. Univ. of Cal. Press, 1961, pp. 57-78.

2. H. CRAMER. Stochastic processes as curves in Hilbert space. *Theory of Probability and Its Applications, IX* (1964), 169-177.

3. M.H.A. DAVIS and P. VARAIYA. The multiplicity of an increasing family of σ-fields. *The Annals of Probability 2* (1974), 958-963.

4. C. DELLACHERIE and P.A. MEYER. *Probabilités et Potentiel.* Publ. de L'institut de Math. de L'Univ. Strasbourg No. XV(1975) and No. XVII (1980).

5. J.L. DOOB. *Stochastic Processes.* Wiley, New York, 1953.

6. B.V. GNEDENKO and A.N. KOLMOGOROV. *Limit Distributions for Sums of Independent Random Variables.* Addison-Wesley, Reading, 1954.

7. R.K. GETOOR. Markov Processes: Ray Processes and Right Processes. Lecture Notes in Math. No. 440. Springer-Verlag, Berlin, 1975.

8. S.W. HE and J.G. WANG. The total continuity of natural filtrations. *Sem. de Prob. XVI,* pp. 348-354. Lecture Notes in Math. No. 920. Springer-Verlag, Berlin, 1981.

9. T. HIDA. Canonical representations of Gaussian processes and their applications. *Memoirs of the College of Science, Univ. of Kyoto Series A,* (1), Vol. 33 (1960), 109-155.

10. F.B. KNIGHT. A reduction of continuous square-integrable martingales to Brownian motion. Lecture Notes in Math. No. 190. Springer-Verlag, Berlin, 1971.

11. F.B. KNIGHT. Essays on the Prediction Process. *Inst. of Math. Statistics Lecture Notes Series No. 1,* 1981.

12. F.B. KNIGHT. A post-predictive view of Gaussian processes. *Ann. Sci. de L'École Normale Superieure, Series 4, Vol. 16* (1983), 541-566.

13. H. KUNITA and S. WATANABE. On square-integrable martingales. *Nagoya Math. J. 30* (1967), 209-245.

14. J. de SAM LAZARO and P.A. MEYER. Questions de la théorie des flots VI. *Sem. de Prob. IX,* pp. 73-88. Lecture Notes in Math. No. 465. Springer-Verlag, Berlin, 1981.

15. Y. LE JAN. Temps d'arrêt stricts et martingales de saut. *Z. Wahrscheinlichkeitstheorie verw. Geb. 44* (1978), 213-225.

16. Y. LE JAN. Martingales et changement de temps. *Sem. de Prob. XIII,* pp. 385-400. Lecture Notes in Math. No. 721. Springer-Verlag, Berlin, 1979.

17. P.A. MEYER. Intégrales stochastiques III. *Sem. de Prob. I.* Lecture Notes in Math. No. 39. Springer-Verlag, Berlin, 1967.

18. P.A. MEYER. Demonstration simplifiée d'un théorème de Knight. *Sem. de Prob. V,* pp. 191-195. Lecture Notes in Math. No. 191. Springer-Verlag, Berlin, 1971.

19. S. WATANABE. On discontinuous additive functionals and Lévy measures of a Markov process. *Jap. J. Math. 34* (1964), 53-79.

20. J. WALSH and P.A. MEYER. Quelques applications des résolvantes de Ray. *Invent. Math. 14* (1971), 143-166.

F. B. KNIGHT
Department of Mathematics,
University of Illinois,
Urbana, IL 61801

*Seminar on Stochastic Processes, 1984*
Birkhäuser, Boston, 1986

# A TIME REVERSAL STUDY OF EXIT/ENTRANCE PROCESSES

by

JOANNA B. MITRO

## 0. Introduction

In [7],[9] Maisonneuve examined "entrance" and "exit" processes associated to a semi-regenerative process $(X;M)$. That is, $M$ is a right closed homogeneous random subset of $\mathbb{R}^+$, and $X$ is $M$-regenerative:

$$E^x(f\circ\theta_T|F_T) = E^{X_T}(f) \text{ a.s. on } \{T < \infty\}$$

for any stopping time $T$ whose graph $[\![T]\!]$ is contained in $M$. For $t \geq 0$, with the conventions $\inf\phi = \infty$ and $\sup\phi = 0$, define

$$D_t = \inf\{s>t:s\in M\}, \quad G_t = \sup\{s\leq t:s\in M\}, \quad g_t = \sup\{s<t:s\in M\}$$

$$R_t = D_t - t, \quad A_t = t - G_t, \quad a_t = t - g_t .$$

Maisonneuve showed that the "entrance process" $(R_t, X_{D_t})$ is strong Markov, and the "exit" process $(A_t, X_{G_t})$ is strong Markov provided $M$ is closed. (The Markov property of $(A_t, X_{G_t})$ requires special assumptions when $M$ is only right closed; see [7] or[9]).

In this paper we consider the entrance process $(R_t, X_{D_t})$ and a modified exit process (or "left-exit process") $(a_t, X_{g_t-})$ which we relate to an entrance process by time reversal. Maisonneuve considered this process in [9], but not from the point of view of time reversal, although he noted that it should be possible to do so. Recent techniques make a time reversal approach more natural and tractable now. The first step in this direction was Maisonneuve's use of "exit-system"-type methods [8] in [7]. Since then the theory of exit systems has been extended to Markov processes on the line, i.e., indexed by $\mathbb{R}$([3],[11]), making Maisonneuve's methods easily coupled with time reversal arguments.

The idea to look at the entrance and left-exit processes as duals arose in connection with dual Markov processes. The most natural (although not necessarily most general) setting for this problem is the following: $(X;M)$ and $(\hat{X};\hat{M})$ are "dual" semi-regenerative processes comprised of dual (strong) Markov processes $X$ and $\hat{X}$ together with "dual" closed random sets $M$ and $\hat{M}$. In this context the problem can be handled using methods of [11].

Duality of the entrance process for $(X;M)$ and the left-exit process for $(\hat{X};\hat{M})$ is not surprising; indeed, it is a rather intuitive consequence of time reversal. Here the result falls out of a simple calculation using a "two-sided Markov process" ([10],[11]). Also, the setting is different from that of [9], and we show directly in this case that the left-exit process is moderate Markov [4].

## 1. Prerequisites

Let $X = (\Omega, F, F_t, X_t, \theta_t, P^x)$ and $\hat{X} = (\hat{\Omega}, \hat{F}, \hat{F}_t, \hat{X}_t, \hat{\theta}_t, \hat{P}^x)$ be standard processes in duality relative to the $\sigma$-finite measure $\xi$

on their common Lusinian state space $(E,\mathcal{E})$. $(P_t)$ and $(\hat{P}_t)$ will denote their respective semigroups. Duality means

$$(1.1)\qquad \int f(x)\hat{P}_t g(x)\xi(dx) = \int g(x)P_t f(x)\xi(dx)$$

for both $f,g \in b\mathcal{E}$, and further that $\xi$ is an excessive reference measure for both processes (see [2]). For simplicity we assume X and $\hat{X}$ have infinite lifetimes a.s. (i.e., they are Hunt processes).

We associate to this dual pair a *two-sided Markov process* as follows. Let W denote the space of r.c.l.l. paths from $\mathbb{R}$ into E, with coordinate maps $Z_t(w) = w(t)$ and shift operators $\sigma_t$:

$$Z_t\circ\sigma_s(w) = Z_{t+s}(w)$$

Define, for $t \in \mathbb{R}$,

$$(1.2)\qquad \hat{Z}_t = Z_{(-t)-},\ \hat{\sigma}_t = \sigma_{-t},$$

and let $(\mathcal{G}_t^\circ)$ and $(\hat{\mathcal{G}}_t^\circ)$ denote the natural filtrations of Z and $\hat{Z}$ respectively. The $\sigma$- algebra on W generated by $(Z_t,\ t\in\mathbb{R})$ is denoted $\mathcal{G}^\circ$.

The construction in [10] produces a $\sigma$-finite, stationary measure Q on $(W,\mathcal{G}^\circ)$ under which Z is Markov with respect to $(P_t)$ and $\hat{Z}$ is Markov with respect to $(\hat{P}_t)$. The $\sigma$-algebras $\mathcal{G}^\circ,\mathcal{G}_t^\circ,\hat{\mathcal{G}}_t^\circ$ are completed by adjoining the ideal of Q-null sets; the resulting filtrations $(\mathcal{G}_t)$ and $(\hat{\mathcal{G}}_t)$ are right continuous. Projections $\tau_t:W\to\Omega\ (t\in\mathbb{R})$ given by

$$(1.3)\qquad X_s\circ\tau_t(w) = Z_{s+t}(w),$$

(along with the dual conterpart $\hat{\tau}_t$) enable us to relate the two-sided process $(W, G, Q, (Z_t, G_t, \sigma_t), (\hat{Z}_t, \hat{G}_t, \hat{\sigma}_t))$ to the original dual pair. For instance, for $f \in F$ and T a $(G_t)$ - stopping time:

$$(1.4) \qquad Q(f \circ \tau_T | G_T) = P^{Z_T}(f).$$

We shall also use the notation

$$(1.5) \qquad \tilde{\tau}_t = \hat{\tau}_{-t} ,$$

and $\mathbb{R}^{++} = (0,\infty)$.

Let $M \subset \mathbb{R}^{++} \otimes \Omega$ be a closed homogeneous $(\bar{F}_t)$- optional random set. Set $R = \inf\{t: t \in M\}$ and let $({}^*P, B)$ denote the (optional) exit system [8] for (X,M). (The additive functional B will be identified with the random measure having B as distribution function.) In [11] we showed that there is a corresponding random set N $\mathbb{R} \times W$ and exit system $({}^*P, \Lambda)$ for Z satisfying

$$(N-t) \cap \mathbb{R}^{++} = M \circ \tau_t ,$$

$$\Lambda(\omega, t+A) = B(\tau_t\, \omega, A) \quad \text{for } A \in \mathcal{B}(\mathbb{R}^+),$$

and

$$(1.6) \qquad Q \sum_{t \in N_\ell} Y_t \cdot H_t \circ \tau_t = Q \int Y_t \, {}^*P^{Z_t}(H_t) \Lambda(dt)$$

for $Y \geq 0$ and $(G_t)$-optional and $H \in (\mathcal{B}(\mathbb{R}) \otimes F^*)^+$, where $N_\ell$ denotes the set of left endpoints of intervals contiguous to N.

Under our duality hypothesis the structure of M takes a very specific form [5] which enables us to define a dual set $\hat{M}$ for $\hat{X}$. In

[11] we cast the corresponding exit system for $\hat{Z}$ as a "co-exit" system for Z; here we will need the analogous result for the *$\hat{Z}$-predictable* exit system $({}^{\#}\hat{P},\hat{\Gamma})$ obtained from the predictable exit system $({}^{\#}\hat{P},\hat{C})$ for $\hat{X}$ [5]:

$$(1.7) \qquad Q \sum_{t \in N_r} V_s \cdot K_s \circ \tilde{\tau}_s = Q \int V_s \, {}^{\#}\tilde{P}^{Z_{s-}}(K_s) \tilde{\Gamma}(ds)$$

for $K \in (\mathcal{B}(\mathbb{R}) \otimes \hat{F}^*)^+$ and $V$ "co-predictable", i.e., $t \to V_{-t}$ is $(\hat{\mathcal{G}}_t)$-predictable. Here $N_r$ is the set of right endpoints of intervals contiguous to $N$ and $\tilde{\Gamma}(ds) = \hat{\Gamma}(-ds)$. $\hat{C}$ is a natural additive functional; its Revuz measure $\mu$ satisfies [1]

$$(1.8) \qquad Q \int f(s,Z_s) \tilde{\Gamma}(ds) = \iint f(s,x) \mu(dx) ds.$$

## 2. Duality of Entrance and Left-Exit Processes

In this section we verify that the process $Y_t = (R_t, X_{D_t})$ is dual to $\tilde{Y} = (\hat{a}_t, \hat{X}_{\hat{g}_{t-}})$, the left-exit process associated to $(\hat{X};\hat{M})$. What we actually do is identify the duality measure and show that the transition functions of $Y$ and $\tilde{Y}$ satisfy the appropriate analogue of (1.1). (These processes are technically in *weak* duality: neither process has a reference measure in general.)

We first recall some facts about $Y$ from [7]. Set $\overline{E} = \mathbb{R}^+ \times E$, $\Delta = (+\infty, \delta)$ and $\overline{E}_\Delta = E \cup \{\Delta\}$ ($\Delta$ is adjoined as an isolated point). Define $P^\delta = \varepsilon_{[\delta]}$, where $X_t[\delta] = \delta$ all t, and $X_\infty(\omega) = \delta$ for all $\omega \in \Omega$. The family of kernels $(\overline{P}_t)$ on $(\overline{E}, \overline{E}^*_\Delta)$ defined by

$$\overline{P}_t((r,x),f) = f(r-t,x) \quad \text{if} \quad t < r$$

(2.1)

$$= E^x(f \circ Y_{t-r}) \text{ if } t \geq r$$

is a semigroup for Y. This process is strong Markov with respect to the family $(\overline{F}_t) = (F_{D_t})$ and the semigroup $(\overline{P}_t)$. (Note that points $(0,x)$ for $x \notin F = \mathrm{reg}(R)$ are branch points. Glover [6] has shown that Y is a right process when restricted to the set of non-branch points.)

The left-entrance process can be analyzed by the methods of [7]. We state the results for $\check{Y}_t = (a_t, X_{g_t-})$ rather than $\tilde{Y}$ to simplify the notation. Dual formulas will hold for $\tilde{Y}$.

Set $\check{F}_t = F_{g_t-}$. (Recall that $F \in F$ is $\check{F}_t$- measurable iff $F = V_{g_t}$ on $\{g_t < \infty\}$ for some predictable process V.) Define, for $h \in F^*$, $x \in E$:

$$H((a,x),h) = \frac{{}^{\#}P^x(h(R-a,X_R);a<R)}{{}^{\#}P^x(a<R)} \quad \text{for } a>0 \quad (\tfrac{c}{0}=0)$$

(2.2)

$$= E^x(h(R,X_R)) \qquad a=0.$$

Then the following results are valid, and may be proved like their counterparts in [7]:

(2.3) PROPOSITION. *For any predictable* $\check{F}_t$- *stopping time* T *and* $h \in bF^*$,

(i) $H(\check{Y}_T,1) > 0$ *a.s. on* $\{g_T < T < \infty\}$ ,

(ii) $E^\mu(h \circ Y_T | \check{F}_{T-}) = H(\check{Y}_T,h)$ *a.s. on* $\{T < \infty\}$ .

(2.4) THEOREM. *Assume that almost surely* M *has no isolated* *points. Then* $(\check{\mathcal{F}}_t)$ *is a filtration and the process* $(\check{Y}_t)$ *is moderate Markov with respect to* $(\check{\mathcal{F}}_t)$ *and the semigroup* $(\check{P}_t)$ *of kernels on* $(\overline{E},\mathcal{E}^*)$ *given by*

$$(2.5)\qquad \check{P}_t((a,x),f) = f(a+t,x)\cdot H((a,x),(t,\infty)\times E) + H((a,x),\check{f}_t),$$

where $\check{f}_t$ is the function defined by

$$(2.6)\qquad \check{f}_t(r,y) = E^y(f\circ \check{Y}_{t-r}) \quad \text{if } r \le t$$
$$= 0 \qquad \text{if } r > t.$$

REMARKS. The proof of (2.3) (ii) relies in part (on the set $\{g_T = T\}$) on the moderate Markov property of the (Hunt) process X, [4]. The filtration property of $(\check{\mathcal{F}}_t)$ is proved in [5]. Assuming M has no isolated points, we can extend (2.3) to general $(\check{\mathcal{F}}_t)$ stopping times T, in which case the equality in (2.3) (ii) holds only a.s. on $\{g_t < T < \infty\}$.

From now on, we make the assumption that

almost surely, M has no isolated points.

By time reversal, the same holds for $\hat{M}$. The main result of this section is

(2.7) THEOREM. *The processes* $\check{Y}$ *and* $Y$ *are in weak duality relative to the measure*

$$\nu(dx,ds) = \eta(dx,ds) + (1_{F \cap \hat{F}}(x)\xi(dx)) \times \varepsilon_{\{0\}}(ds)$$

where $\hat{F} = \text{reg}(\hat{R})$ and

$$\eta(dx,ds) = {}^{\#}\hat{P}^{x}(\hat{R} > s)1_{(0,\infty)}(s)\mu(dx)ds.$$

PROOF. We need to show that, for $f,g \in (\mathcal{B}(\mathbb{R}^{+})\otimes E)^{+}$ and $t > 0$,

(2.8) $$\int f(s,x)\overline{P}_{t}g(s,x)\nu(ds,dx) = \int g(s,t)\tilde{P}_{t}f(s,x)\nu(ds,dx),$$

where $(\tilde{P}_{t})$ is the dual counterpart to $(\check{P}_{t})$. Note that f and g vanish at $\Delta$. We proceed by computing

(2.9) $$Q[(f(Y_{0})g(Y_{t}))\circ\tau_{0}].$$

Rewriting (2.9) in terms of N and Z yields three terms: one in case $0 \in N-N_{\ell}$,

(2.10) $$Q[f(0,Z_{0})\{g(Y_{0})1_{\{R=0\}})\circ\tau_{0}] ;$$

a second case 0 and t lie in the same component of $N^{c}$,

(2.11) $$Q\sum_{\substack{u\in N^{r}\\ u>t}} f(u,Z_{u})g(u-t,Z_{u})(1_{\{u<\hat{R}\}}\circ\tilde{\tau}_{u});$$

and the third for t outside the component of $N^{c}$ containing 0,

(2.12) $$Q \sum_{\substack{u \in N_r \\ 0<u<t}} f(u,Z_u)(1_{\{u<\hat{R}\}}\circ\tilde{\tau}_u)(g(Y_{t-u})\circ\tau_u).$$

Note that $Q(s \in N_\ell) = Q(s \in N_r) = 0$ for any s (see [11]). Using the Markov property (1.4) at 0, and the fact that $\xi(F\Delta\hat{F}) = 0$, (2.10) becomes

(2.13) $$Q(f(0,Z_0)P^{Z_0}(g(Y_t)1_{\{R=0\}})) = \int_{F\cap\hat{F}} f(0,x)P^x(g(Y_t))\xi(dx).$$

Applying (1.7) transforms (2.11) into

(2.14) $$Q\int_t^\infty f(s,Z_s)g(s-t,Z_s)\,{}^{\#}\hat{P}^{Z_{s-}}(s<\hat{R})\tilde{\Gamma}(ds)$$

$$= \int_t^\infty \int_E f(s,x)g(s-t,x)\,{}^{\#}\hat{P}^x(s<\hat{R})\mu(dx)ds.$$

For (2.12), use the fact that $N_r$ is optional with countable sections. Therefore $N_r = \bigcup_n [\![ S_n ]\!]$ where $\{S_n\}$ are pairwise disjoint $(\mathcal{G}_t)$-stopping times, and (2.12) equals

$$Q \sum_n f(S_n,Z_{S_n})1_{\{0<S_n<t\}}1_{\{S_n<\hat{R}\circ\tilde{\tau}_{S_n}\}}g(Y_{t-S_n})\circ\tau_{S_n}.$$

Now apply the strong Markov property in the form

$$Q(h(w,T(w),\tau_T(w))\,|\,\mathcal{G}_T) = \int h(w,T(w),\omega)P^{Z_T(w)}(d\omega) \qquad Q\text{-a.e.}$$

for T a $(\mathcal{G}_T)$-stopping time and $h: W \times \mathbb{R} \times \Omega \to \mathbb{R}^+$ a $(\mathcal{G}_T \otimes \mathcal{B}(\mathbb{R}) \otimes \mathcal{F})$-measurable function, choosing $h(w,s,\omega) = g(Y_{t-s}(\omega))$. This yields

$$Q \sum_{s\in N_{\mathcal{R}},0<s<t} f(s,Z_s)1_{\{s<\hat{R}\}}\circ\tilde{\tau}_s P^{Z_s}(g(Y_{t-s})).$$

Although $P^{Z_s}(g(Y_{t-s}))$ need not be co-predictable, (1.7) holds for $V_s = f(s,Z_s)P^{Z_s}(g(Y_{t-s}))$ by completion, and (2.12) becomes

$$(2.15) \qquad Q\int_0^t f(s,Z_s)P^{Z_s}(g(Y_{t-s}))\,{}^{\#}\hat{P}^{Z_{s-}}(\hat{R}>s)\tilde{\Gamma}(ds)$$

$$= \int_0^t \int f(s,x)P^x(g(Y_{t-s}))\,{}^{\#}\hat{P}^x(\hat{R}>s)\mu(dx)ds.$$

Combining (2.13), (2.14), and (2.15) shows that (2.9) equals

$$\int f(s,x)\overline{P}_t g(s,x)\nu(ds,dx).$$

This gives one half of (2.8). To get the other half, we evaluate (2.9) in another way. Instead of considering the interval of $N^c$ containing 0, we focus on that interval containing t. In this way we obtain three new terms, analogous to the three we had before. The first is for $t \in N°$, the interior of N:

$$(2.16) \qquad Q[g(0,Z_t)(f(\tilde{Y}_t)1_{(\hat{R}=0)})\circ\tilde{\tau}_t]$$

$$= Q(g(0,Z_t)\hat{P}^{Z_t}(f(\tilde{Y}_t);\hat{R}=0))$$

$$= \int_{\hat{F}\cap F} g(0,x)\hat{P}^x(f(\tilde{Y}_t))\xi(dx).$$

For 0 and t in the same component of $N^c$ we have

$$Q \sum_{\substack{s\in N_{\mathcal{R}}\\ t<s<\infty}} g(s-t,Z_s)\, f(s,Z_s)1_{\{\hat{R}>s\}}\circ\tilde{\tau}_s$$

$$= \int_t^\infty \int g(u-t,x)f(u,x)\,{}^{\#}\hat{P}^x(\hat{R}>u)\mu(dx)du$$

(2.17) $$= \int\int g(s,x)f(s+t,x)\frac{{}^{\#}\hat{P}^x(\hat{R}>s+t)}{{}^{\#}\hat{P}^x(\hat{R}>s)}\eta(dx,ds).$$

Finally, in case 0 lies outside the component of $N^c$ containing t:

$$Q\sum_{\substack{s\in N_{\mathcal{R}}\\ t<s<\infty}} g(s-t,Z_s)(f(\tilde{Y}_s)1_{\{s-t<\hat{R}<s\}})\circ\tilde{\tau}_s$$

(2.18) $$= \int_t^{\infty}\int g(u-t,x)\,{}^{\#}\hat{P}^x(f(\tilde{Y}_u);u-t<\hat{R}<u)\,\mu(dx)\,du.$$

The calculations leading to (2.17) and (2.18) use (1.7), rewritten in terms of Revuz measure. Because M has no isolated points, $\hat{R}<u$ implies $g_u>\hat{R}$ and hence $f(\tilde{Y}_u)=f(Y_{u-\hat{R}})\circ\theta_{\hat{R}}$. Using the strong Markov property of $\hat{X}$ under ${}^{\#}\hat{P}^x$ at the stopping time $\hat{R}$, (2.18) becomes (using notation dual to (2.6))

(2.19) $$\int_t^{\infty}\int g(u-t,x)\,{}^{\#}\hat{P}^x(\tilde{f}_u(\hat{R},\hat{X}_{\hat{R}});u-t<\hat{R})\,\mu(dx)\,du$$

$$= \int_0^{\infty}\int g(s,x)\,\frac{{}^{\#}\hat{P}^x(\tilde{f}_{s+t}(\hat{R},\hat{X}_{\hat{R}}),s<\hat{R})}{{}^{\#}\hat{P}^x(s<\hat{R})}\,\eta(dx,ds).$$

Combining (2.16),(2.17) and (2.19) gives the right-hand side of (2.8). This finishes the proof.

REMARK. In the special case that $M=\{(t,w): X_t\in E_0\}^-$ (where "−" denotes closure in $]\!]0,\infty[\![$) for $E_0$ a closed, finely perfect subset of E which is *left jump free for* $\hat{X}$ (equivalently, right jump free for X) (see [5], section 10), the right continuous modification of $\tilde{Y}$ is $(\hat{A}_t,\hat{X}_{\hat{G}_t})$, which is strong Markov. Thus, at least in this case, the entrance process Y possesses a *strong* Markov dual.

## References


1. B.W. ATKINSON and J.B. MITRO. Applications of Revuz and Palm type measures for additive functionals in weak duality. *Seminar on Stochastic Processes* (1982), pp.23-49. Birkhäuser, Boston, 1982.

2. R.M. BLUMENTHAL and R.K. GETOOR. *Markov Processes and Potential Theory.* Academic Press, New York, 1968.

3. BOUTABIA. Thèse de troisième cycle (unpublished).

4. K.L. CHUNG. On the fundamental hypotheses of Hunt processes, Instituto di Alta Matematica, *Symposia Mathematica* IX (1972), 43-52.

5. R.K. GETOOR and M.J. SHARPE. Excursions of dual processes. *Advances in Mathematics* 45 (1982), 259-309.

6. J. GLOVER. Discontinuous time changes of semiregenerative processes and balayage theorems. *Z. Wahrscheinlichkeits-theorie verw. Gebiete 65* (1983), 145-160.

7. B. MAISONNEUVE. Entrance-Exit results for semi-regenerative processes. *Z. Wahrscheinlichkeitstheorie verw. Gebiete* 32 (1975), 81-94.

8. B. MAISONNEUVE. Exit systems. *Ann. Probability* 3 (1975), 399-411.

9. B. MAISONNEUVE. Systèmes régénératifs. *Asterisque, No.* 15. *Soc. Math.* France, Paris, 1974.

10. J.B. MITRO. Dual Markov processes: construction of a useful auxiliary process. *Z. Wahrscheinlichkeitstheorie verw. Gebiete* 47 (1979), 139-156.

11. J.B. MITRO. Dual exit systems for dual Markov processes. *Z Wahrscheinlichkeitstheorie verw. Gebiete* 66 (1984), 259-267.


J.B. MITRO
Department of Mathematics
University of Cincinnati
Cincinnati, Ohio 45221

*Seminar on Stochastic Processes, 1984*
Birkhäuser, Boston, 1986

# ON THE CONTINUITY OF THE LOCAL TIME OF STABLE PROCESSES

by

EDWIN PERKINS

## 1. Introduction and Statement of Result

Let $X(t)$ denote a strictly stable process of index $\alpha > 1$. That is, $X(0) = 0$, $X$ has stationary independent increments, and

$$E(e^{izX(t)}) = \exp\{-t\psi(z)\},$$
$$\psi(z) = c_0|z|^\alpha(1 - ih\ \mathrm{sgn}(z)\tan(\pi\alpha/2)). \tag{1}$$

Here $c_0 > 0$, $\alpha \in (1,2]$, and $|h| \le 1$. It is well-known (see [3], [8]) that $X$ has a jointly continuous local time, $L_t^x$, which we may normalize so that

$$\int_B L_t^x\,dx = \int_0^t I_B(X_s)ds \qquad \text{for all Borel } B \subset \mathbb{R} \text{ and } t \ge 0. \tag{2}$$

In this note we find the exact modulus of continuity of $L_\cdot^x$ uniformly in x. This problem arose in [2] in the course of proving that the appropriate Hausdorff measure of the level sets of X coincides with the local time of X for all levels simultaneously. Indeed the problem was solved for Brownian motion in [11] for precisely this reason.

The proof is a little more involved than one might initially suspect and in fact relies on a recent result of Martin Barlow [1] concerning the modulus of continuity of $L_t^{\cdot}$ uniformly in t. The well-known lemma of Garsia, Rodemich and Rumsey [7] is applied to $L_t^{\cdot}$ to obtain the *best constant* in the modulus of continuity for L in the time variable uniformly in space. In order to carry this out, one has to carefully examine the random coefficients that arise when one applies the G.R.R. lemma to $L_t^{\cdot}$ (see Lemma 5).

The method can also be used to answer similar questions about the jointly continuous local time of other Lévy processes, although the information obtained may not be as precise (see Theorem 8).

The exact modulus of continuity of $L_{\cdot}^0$ was obtained by Hawkes [9]. If $\tau(a)$ is the right-continuous inverse of $L_t^0$,

$$(3) \qquad \gamma = 1 - 1/\alpha,$$

and $\rho$ is given by

$$(4) \qquad \rho^{-\gamma} = \pi^{-1}\,\Gamma(1+1/\alpha)\Gamma(1-1/\alpha)c_0^{-1/\alpha}\,\mathrm{Re}[(1-ih\tan(\pi\alpha/2))^{-1/\alpha}],$$

then $T(t) = \rho^{-1}\tau(t)$ is a stable subordinator of index $\gamma$, scaled so that $E(e^{-\lambda T(t)}) = e^{-t\lambda^{\gamma}}$ (see [9]). Moreover (Lemma 1 of [9]),

$$(5) \qquad P(T(1) \le x) \sim c_1\, x^{\gamma/2(1-\gamma)} \exp\{-c_2\, x^{-\gamma/(1-\gamma)}\} \quad \text{as } x \downarrow 0,$$

where $f(x) \sim g(x)$ means their ratio converges to one, $c_1 > 0$, and

$$(6) \qquad c_2 = (1-\gamma)\gamma^{\gamma/(1-\gamma)}.$$

Hawkes (see also the remarks in [5]) used this estimate to show

$$(7) \qquad \lim_{s\to 0^+} \sup_{t\le 1} (L_{t+s}^0 - L_t^0)\,\phi_\alpha(s)^{-1} = \rho^{-\gamma} c_2^{\gamma-1}\,\gamma^{1-\gamma}$$

where

$$(8) \qquad \phi_\alpha(u) = u^{1-1/\alpha}(\log 1/u)^{1/\alpha} \quad \text{for } u \in (0,1]$$

and $\gamma$, $c_2$ and $\rho$ are given by (3), (4) and (6) throughout this paper.

We are ready to state our main result. The proof is given in section 3.

THEOREM 1. *Let* $L_t^x$ *denote the jointly continuous local time of the strictly stable process* X *of index* $\alpha > 1$ *whose characteristic function is given by* (1). *Assume* L *is normalized so that* (2) *holds. Then*

$$\lim_{s\to 0^+} \sup_{t\le 1, a\in \mathbb{R}} (L_{t+s}^a - L_t^a)\phi_\alpha(s)^{-1}$$

(9)

$$= \lim_{s\to 0^+} \sup_{t\le 1} (L_{t+s}^{X(t)} - L_t^{X(t)})\phi_\alpha(s)^{-1} = \theta_0,$$

*where* $\theta_0 = \rho^{-\gamma} c_2^{\gamma-1}$ $(= \rho^{-\gamma}\alpha^{1/\alpha}(1-\alpha^{-1})^{\alpha^{-1}-1})$, $\rho^{-\gamma}$ *is given by* (6), *and* $\phi_\alpha$ *is given by* (8). □

In particular (9) always is greater than (7), and their ratio goes to $\infty$ as $\alpha \to 1$.

For Brownian motion the theorem states

$$\lim_{s\to 0^+} \sup_{t\le 1, a\in \mathbb{R}} (L_{t+s}^a - L_t^a)(s \log 1/s)^{-\frac{1}{2}} = \sqrt{2}.$$

This was proved in [11, Lemma 5(c)] but the argument given there used the Ray-Knight theorems on the Markov properties of Brownian local time and therefore does not extend to the present situation.

Throughout, c denotes a generic positive constant whose value may change from line to line.

## 2. Preliminary Lemmas

*Notation.* If $M, d > 0$, let

$$S(d,M) = \{kd \mid k \in \mathbb{Z},\ |kd| \le M\}$$

$$S^+(d,M) = S(d,M) \cap [0,\infty)$$

$$S(d) = S(d,\infty),\quad S^+(d) = S^+(d,\infty).$$

LEMMA 2. *Fix* $u \in (0,1)$ *and suppose* $q,\theta > 0$ *satisfy*

$$\max(1, q+1-\alpha^{-1}) < c_2\, \rho^{\alpha-1}\, \theta^{\alpha}. \tag{10}$$

*Then for a.a.* $\omega$ *and large enough* $n$ *(depending on* $\omega$*),*

$$\sup\{L^x_{t+u^n} - L^x_t \mid t \in S^+(u^n n^{-1},1),\ x \in S(u^{nq})\} \le \theta\phi_\alpha(u^n). \tag{11}$$

PROOF: If $L(n)$ denotes the supremum on the left side of (11), then

$$P(L(n) > \theta\phi_\alpha(u^n)) \le \Sigma_{t\in S^+(u^n n^{-1},1)} \Sigma_{x\in S(u^{nq})} P(X(s) = x$$

$$\exists\ s \in [t, t+u^n]) \times P(L^0_{u^n} > \theta\phi_\alpha(u^n))$$

$$\le P(\tau(\theta\phi_\alpha(u^n)) \le u^n)\Sigma_{t\in S^+(u^n n^{-1},1)}(2E(\sup_{s\in[t,t+u^n]}|X(s)-X(t)|)u^{-nq}+1)$$

$$\le P(T(\theta\phi_\alpha(u^n)) \le u^n\rho^{-1})(nu^{-n}+1)(2u^{-n(q-1/\alpha)}E(\sup_{t\le 1}|X(t)|)+1).$$

As $X(t) - tE(X(1))$ is a martingale in $L^p$ for $p < \alpha$, $E(\sup_{t\le 1}|X(t)|) < \infty$ and so the above is bounded by

$$cn(u^{-n(q+1-\alpha^{-1})} + u^{-n})P(T(1) < \theta^{-1/\gamma}\,\phi_\alpha(u^n)^{-1/\gamma}\,u^n\rho^{-1})$$

$$\le cn(u^{-n(q+1-\alpha^{-1})} + u^{-n})n^{-\frac{1}{2}}\,u^{nc_2\rho^{\alpha-1}\theta^{\alpha}} \qquad \text{(by (5))}.$$

As this is summable over $n$ if (10) holds, the result follows from the

Borel-Cantelli Lemma. □

In [1] Barlow refines results of Getoor and Kesten [8] and obtains what appear to be fairly precise estimates on the modulus of continuity in the space variable of the local time of a Lévy process. The fundamental estimate we need follows from Lemmas 2.8, 2.4 and the remarks in section 4.2 of [1]:

LEMMA 3. *There is a* $c_3 = c_3(\alpha,h,c_0)$ *such that*

$$P(|L_t^a \wedge \lambda - L_t^b \wedge \lambda| > x) \le 2\exp\{-c_3 x^2/(\lambda(b-a)^{\alpha-1})\}$$

*for all* $\lambda, x > 0$ *and real* $a,b$ *satisfying* $|a-b| \le 1$. □

This estimate is used in the following result of Garsia, Rodemich and Rumsey [7] (see [6]).

LEMMA 4. *Let* $p$ *and* $\psi$ *be strictly increasing functions on* $[0,\infty)$ *such that* $p(0) = \psi(0) = 0$, $\lim_{t\to\infty} \psi(t) = \infty$ *and* $\psi$ *is convex. Let* $f$ *be a continuous function on an interval* $I$ *such that*

$$\int_I \int_I \psi\left(\frac{|f(x)-f(y)|}{p(|x-y|)}\right) dx\, dy \le \Gamma < \infty.$$

*Then, for all* $x,y \in I$,

$$|f(x) - f(y)| \le 8 \int_0^{|x-y|} \psi^{-1}(\Gamma u^{-2})\, dp(u).$$ □

By setting $\psi(x) = e^{x^2} - 1$, $p(u) = (2\lambda/c)^{\frac{1}{2}} u^{(\alpha-1)/2}$ and using Lemma 3 in the above, Barlow is able to show

(12) $\exists\, c_4 = c_4(\alpha,h,c_0)$ *such that for a.a.* $\omega$, *and all* $t > 0$

$$\exists\ \varepsilon_t(\omega) > 0 \text{ such that } |L_s^a - L_s^b| \le c_4 (\sup_x L_t^x)^{\frac{1}{2}} |b-a|^{\frac{\alpha-1}{2}} (\log \frac{1}{|b-a|})^{\frac{1}{2}}$$

*for all* $s \in [0,t]$ *and* $|b-a| < \varepsilon_t(\omega)$.

Although these results are only explicitly given in [1] for symmetric stable processes, the same computations go through in the asymmetric case. (12) refines earlier estimates by reducing the power on the logarithm and introducing the factor of $(\sup_x L_t^x)^{\frac{1}{2}}$. It is this latter improvement that we will use.

LEMMA 5. *If* $u \in (0,1)$ *and* $q > 0$, *there is a* $c_5 = c_5(\alpha,q) > 0$ *and an* $N(\omega)$ *such that* $N(\omega) < \infty$ *a.s., and for* $n \ge N$, $t \in S^+(u^n n^{-1},1)$ *and* $|a-b| \le u^{nq}$,

$$(13)\quad \sup_{s \le u^n} |(L_{t+s}^a - L_t^a) - (L_{t+s}^b - L_t^b)|$$

$$\le c_5 (\sup_x (L_{t+u^n}^x - L_t^x) + u^n)^{\frac{1}{2}} |b-a|^{(\alpha-1)/2} (\log(|b-a|^{-1}))^{\frac{1}{2}}.$$

PROOF: Choose $p > (2/q) \vee 1$ and then $r > 0$ so that $c_3(rp)^{-1} > 1$. Let $\psi_r(x) = e^{rx^2} - 1$, $p(u) = \lambda^{\frac{1}{2}} u^{(\alpha-1)/2}$ and for each bounded interval I, $t \ge 0$, and $\lambda > 0$ define

$$\Gamma(I,\lambda,t) = \int_I \int_I \psi_r(\sup_s |L_s^a(X^{(t)}) \wedge \lambda - L_s^b(X^{(t)}) \wedge \lambda | p(|b-a|)^{-1}) da db.$$

Here $L_s^a(X^{(t)})$ denotes the local time of the stable process $X^{(t)}(s) = X(t+s) - X(t)$. As $\psi_r^{-1}(y) = (\log(y+1))^{\frac{1}{2}} r^{-\frac{1}{2}}$, Lemma 4 implies that for $s \ge 0$ and $a,b \in I$,

$$|L_s^a(X^{(t)}) \wedge \lambda - L_s^b(X^{(t)}) \wedge \lambda|$$

$$\le 8(\lambda/r)^{\frac{1}{2}} \int_0^{|b-a|} (\log(\Gamma v^{-2} + 1))^{\frac{1}{2}} d(v^{(\alpha-1)/2}) \qquad (\Gamma = \Gamma(I,\lambda,t))$$

$$\le 8(\lambda/r)^{\frac{1}{2}}\Big[\int_0^{|b-a|} \log^+\Big(\frac{\Gamma+v^2}{1+v^2}\Big)^{\frac{1}{2}}\, d(v^{(\alpha-1)/2})$$

$$+ \int_0^{|b-a|} \Big(\log\Big(\frac{1+v^2}{v^2}\Big)\Big)^{\frac{1}{2}}\, d(v^{(\alpha-1)/2})\Big]$$

$$\le c_6\lambda^{\frac{1}{2}}[(\log^+\Gamma)^{\frac{1}{2}} + (\log(|b-a|^{-1}))^{\frac{1}{2}}]|b-a|^{(\alpha-1)/2} \quad \text{for } |b-a| \le \tfrac{1}{2},$$

where we have integrated the second integral by parts. Here $\log^+ x = \log(x \vee 1)$ and $c_6$ depends only on $\alpha$ and $r$ and hence ultimately on $\alpha$ and $q$. Note that $L_s^a(X^{(t)}) = L_{s+t}^{a+X(t)} - L_t^{a+X(t)}$, so we have shown that for any $a,b \in I$ with $|b-a| \le \frac{1}{2}$ and any $s,t \ge 0$, $\lambda > 0$,

$$(14) \quad |(L_{s+t}^a - L_t^a)\wedge\lambda - (L_{s+t}^b - L_t^b)\wedge\lambda|$$

$$\le c_6\lambda^{\frac{1}{2}}[(\log^+\Gamma(I-X(t),\lambda,t))^{\frac{1}{2}} + \log(|b-a|^{-1})|b-a|^{(\alpha-1)/2},$$

where $I - X(t) = \{x - X(t) \mid x \in I\}$.

Some estimates on the size of $\Gamma(I,\lambda,t)$ are needed. Let I be an interval of length one.

$$E(\Gamma(I,\lambda,t)^p) = E(\Gamma(I,\lambda,0)^p)$$

$$\le \int_I\int_I E(\psi_{pr}(\sup_s|L_s^a\wedge\lambda - L_s^b\wedge\lambda|p(|b-a|)^{-1}))dadb$$

$$= \int_I\int_I\int_0^\infty P(\sup_s|L_s^a\wedge\lambda - L_s^b\wedge\lambda| \ge p(|b-a|)(rp)^{-\frac{1}{2}}(\log(y+1))^{\frac{1}{2}})dydadb$$

$$\le \int_0^\infty 2(y+1)^{-c_3/rp}dy \quad \text{(by Lemma 3)}$$

$$= c(\alpha,h,c_0,q) < \infty$$

by the choice of $r$. Let $\mathcal{I}_n = \{[x,x+1] \mid 2x \in \{-2n,-2n+1,\dots,2n\}\}$ and

$$\hat{\Gamma} = \sup\{\Gamma(I,\lambda,t): I \in \mathcal{I}_n,\ \lambda \in S^+(u^n,n),\ t \in S^+(u^n n^{-1},1)\}.$$

Then

$$P(\hat{\Gamma} \geq u^{-nq}) \leq c\,n^3 u^{-2n} u^{npq}$$

by Chebychev's inequality and the above $L^p$ bound. As this is summable by the choice of $p$, there is an a.s. finite $N(\omega)$ such that if $n \geq N(\omega)$, then

$$\hat{\Gamma} < u^{-nq}. \tag{15}$$

We may also assume $N(\omega)/2 \geq \sup_{t\leq 2} |X(t)|+1$, $u^{Nq} \leq \frac{1}{2}$ and $\sup_{x,t\leq 1} (L^x_{t+u^n} - L^x_t) \leq N$. Fix $n \geq N$, $t \in S^+(u^n n^{-1},1)$ and $a,b$ such that $|a-b| \leq u^{nq}$. Choose $\lambda \in S^+(u^n,n)$ so that $0 \leq \lambda - \sup_x (L^x_{t+u^n} - L^x_t) \leq u^n$. We may assume $a,b \in [-N/2,N/2]$ as (13) is trivial otherwise. If $I$ is the interval with end points $a$ and $b$, then $I - X(t) \subset [-N,N]$ and so $I - X(t) \subset J$ for some $J \in \mathcal{J}_n$. Therefore $\Gamma(I - X(t),\lambda,t) \leq \Gamma(J,\lambda,t)$ and so by substituting (15) into (14) one gets (13) with $c_5 = 2c_6$. □

## 3. Proof of Theorem 1

Fix $u \in (0,1)$ and choose

$$\theta > (\alpha^{-1}+1)^{\alpha^{-1}} c_2^{-\alpha^{-1}} \rho^{\alpha^{-1}-1} \equiv \theta_1 .$$

Then $\theta^\alpha \rho^{\alpha-1} c_2 > 1+\alpha^{-1}$ and so we may choose $q > 2/\alpha$ so that (10) holds. By Lemma 2 and (12) there is an a.s. finite $N(\omega)$ such that if $n \geq N$, then (11) is true and

$$\sup_{t\leq 2} |L^b_t - L^a_t| \leq c_4(\sup_x L^x_2)^{\frac{1}{2}} |b-a|^{\frac{\alpha-1}{2}} (\log \frac{1}{|b-a|})^{\frac{1}{2}}, \quad \forall\ |a-b| \leq u^{nq}.$$

Fix such an $n$ and for $a \in \mathbb{R}$ choose $b \in S(u^{nq})$ so that $|a-b| \leq u^{nq}$. If $t \in S^+(u^n n^{-1},1)$, then

$$L^a_{t+u^n} - L^a_t \le |L^a_{t+u^n} - L^b_{t+u^n}| + (L^b_{t+u^n} - L^b_t) + |L^b_t - L^a_t|$$

$$\le 2c_4 (\sup_x L^x_2)^{\frac{1}{2}} u^{nq(\alpha-1)/2} (\log u^{-nq})^{\frac{1}{2}} + \theta\phi_\alpha(u^n).$$

The choice of $q$ guarantees that $q(\alpha-1)/2 > \gamma$ and so the right side of the above behaves asymptotically like $\theta\phi_\alpha(u^n)$. We therefore have shown that for a.a. $\omega$ and all $\theta > \theta_1$, there is an $N(\theta,\omega) \in \mathbb{N}$ such that

$$\sup_{t\in S^+(u^n n^{-1},1), a \in \mathbb{R}} L^a_{t+u^n} - L^a_t \le \theta\phi_\alpha(u^n) \tag{16}$$

for $n \ge N$. Unfortunately a quick computation shows that if $X$ is a Brownian motion, $\theta_1 = \sqrt{3}$ ($> \sqrt{2}$!) and so $\theta_1$ is not the best constant. At this point one must take advantage of the factor $(\sup_x L^x_t)^{\frac{1}{2}}$ appearing in (12).

Choose $\theta > \theta_0$, where $\theta_0$ is as in the statement of Theorem 1. Then $\theta^\alpha c_2 \rho^{\alpha-1} > 1$ and so there is a $q > \alpha^{-1}$ satisfying (10). Fix such a $q$ and let $N = N(\theta,\omega)$ be a.s. finite but such that (11), (13) and (16) (the latter with $\theta = \theta_1 + 1$) hold for $n \ge N$. Fix $n \ge N$. If $t \in S^+(u^n n^{-1},1)$, $a \in \mathbb{R}$ and $b \in S(u^{nq})$ satisfies $|b-a| \le u^{nq}$, then combine (11) and (13) to conclude

$$L^a_{t+u^n} - L^a_t \le c_5(\sup_x (L^x_{t+u^n} - L^x_t) + u^n)^{\frac{1}{2}} u^{nq(\alpha-1)/2}(\log u^{-nq})^{\frac{1}{2}} + \theta\phi_\alpha(u^n)$$

$$\le c_5((\theta_1+1)\phi_\alpha(u^n) + u^n)^{\frac{1}{2}} u^{nq(\alpha-1)/2}(\log u^{-nq})^{\frac{1}{2}} + \theta\phi_\alpha(u^n) \qquad \text{(by (16))}$$

$$\le \phi_\alpha(u^n)[\theta + cu^{\frac{n}{2}(q(\alpha-1)-\gamma)} (\log u^{-n})^{1-(\gamma/2)}].$$

As $q > 1/\alpha$, we have shown

$$\limsup_{n\to\infty} \sup_{t\in S^+(u^n n^{-1},1), a\in\mathbb{R}} (L^a_{t+u^n} - L^a_t)\phi_\alpha(u^n)^{-1} \le \theta_0 \quad \text{a.s.}$$

An elementary interpolation argument now shows

$$\limsup_{s\to 0^+} \sup_{t\le 1, a\in\mathbb{R}} (L^a_{t+s} - L^a_t)\phi_\alpha(s)^{-1} \le \theta_0 \quad \text{a.s.}$$

The lower bound is proved as in [11]. Let $\theta < \theta_0$ and $u \in (0,1)$. Then

$$P(\max_{0\le k<u^{-n}} L^{X(ku^n)}_{(k+1)u^n} - L^{X(ku^n)}_{ku^n} \le \theta\phi_\alpha(u^n)) = P(L^0_{u^n} \le \theta\phi_\alpha(u^n))^{u^{-n}}$$

$$= (1 - P(T(1) \le \rho^{-1}\,\theta^{-1/\gamma}\,(\log u^{-n})^{1-1/\gamma}))^{u^{-n}}.$$

(Recall that $T(t) = \rho^{-1}\,\tau(t)$ where $\tau$ is the right-continuous inverse of $L^0_\cdot$). An easy application of (5) shows that the above probabilities are summable over n. The Borel-Cantelli lemma and an elementary interpolation argument now show

$$\liminf_{s\to 0^+} \sup_{t\le 1} (L^{X(t)}_{t+s} - L^{X(t)}_t)\phi_\alpha(s)^{-1} \ge \theta_0 \quad \text{a.s.}$$

This completes the proof. □

Using the condensation argument in [10] it is easy to see that there is an uncountable dense set, $S(\omega)$, of the second category in $[0,\infty)$ such that

$$\limsup_{s\to 0^+} (L^{X(t)}_{t+s} - L^{X(t)}_t)\phi_\alpha(s)^{-1} = \theta_0 \quad \forall\, t \in S(\omega) \quad \text{a.s.}$$

## 4. Extension to Lévy Processes

To illustrate how these methods extend to the local times of other Lévy processes, consider the symmetric Lévy process $X(t)$ with characteristic function

$$E(e^{i\lambda X(t)}) = e^{-t\psi(z)}$$

(17) $$\psi(z) = -\int_{-1}^{1} (e^{izy} - 1 - \frac{izy}{1+y^2})y^{-2}(\log \frac{1}{|y|})^{\alpha}\, dy.$$

Refining results in [8], Barlow [1] showed that for $\alpha > 2$, X has a jointly continuous local time, $L_t^x$, that satisfies

(18) $$\sup_{t\le T}|L_t^a - L_t^b| \le c(\sup_x L_T^x)^{\frac{1}{2}} (\log(|b-a|^{-1})^{1-(\alpha/2)}$$

$$\text{for } |b-a| \le \varepsilon_T(\omega) \text{ for some } \varepsilon_T(\omega) > 0 \text{ a.s.}$$

As before, L is normalized to satisfy (2). If $\alpha \in (1,2]$, $L_t^{\cdot}$ is unbounded on every interval, and if $\alpha \le 1$, L does not exist (see [1]). Therefore assume $\alpha > 2$. If $\tau(t)$ is the right-continuous inverse of $L_{\cdot}^0$, then ([12, Th. 3.21]) $\tau(t)$ is a subordinator and using estimates from [8] one can easily show $E(e^{-\lambda\tau(t)}) = e^{-tg(\lambda)}$ where $g(\lambda) \sim c_7(\log \lambda)^{\alpha-1}$ as $\lambda \to \infty$. In fact $c_7 = \pi^2(\alpha-1)$. The estimates that take the place of (5) may be found in [4, Lemma 1]. Using their upper bound,

$$P(\tau(t) < a) \le e \exp\{-t(c_7 - \varepsilon)(\log 1/a)^{\alpha-1}\} \text{ for small } a \text{ and } \varepsilon > 0,$$

one can prove the following analogue of Lemma 2.

LEMMA 6. *Assume* $\max(1,q) \le \alpha - 1 - \theta$. *Then for* a.a.$\omega$ *and large enough* n *(depending on* $\omega$*)*

(19) $$\sup\{L^x_{t+e^{-n}} - L^x_t | t \in S^+(e^{-n}n^{-1},1),\ x \in S(\exp(-n^q))\} \le n^{-\theta}. \qquad \square$$

A proof similar to that of Lemma 5 gives

LEMMA 7. $\exists\, c_8 = c_8(\alpha)$ *such that for* a.a.$\omega$ *and large enough* n *(depending on* $\omega$*)*

$$(20) \quad \sup_{s\le e^{-n}} |(L^a_{s+t} - L^a_t) - (L^b_{s+t} - L^b_t)|$$

$$\le c_8 (\sup_x (L^x_{t+e^{-n}} - L^x_t) + e^{-n})^{1/2} (\log(|b-a|^{-1})^{1-\alpha/2}$$

for every $t \in S^+(e^{-n}n^{-1},1)$ and $a,b \in \mathbb{R}$ satisfying $|a-b| \le e^{-n}$. □

THEOREM 8. *Let* $X(t)$ *be the Lévy process whose characteristic function is given by* (17) $(\alpha > 2)$ *and let* $L^x_t$ *denote its jointly continuous local time, normalized so that* (2) *holds. Then for any* $\theta < \alpha - 2$

$$(21) \quad \lim_{s\to 0^+} \sup_{t\le 1, a\in \mathbb{R}} (L^a_{t+s} - L^a_t)(\log s^{-1})^\theta = 0$$

but

$$(22) \quad \liminf_{s\to 0^+} \sup_{t\le 1} (L^{X(t)}_{t+s} - L^{X(t)}_t)(\log s^{-1})^{\alpha-2} \ge c_7^{-1} = \pi^{-2}(\alpha-1)^{-1}.$$

PROOF: The proof of (22) is similar to the argument given in the stable case. Use [4, Lemma 1] in place of (5).

Using (18) and Lemma 6 (just as in the proof of (16)) one sees that if $\theta < (1-\alpha^{-1})(\alpha-2)$ then for a.a.$\omega$ and large enough $n$,

$$(23) \quad \sup_{t\in S^+(e^{-n}n^{-1}), x\in\mathbb{R}} L^x_{t+e^{-n}} - L^x_t \le n^{-\theta}.$$

In particular (23) holds for $\theta = (\alpha-2)/2 \equiv \theta_1$. Inductively define $\theta_{i+1} = \frac{1}{2}(\theta_i + (\alpha-2))$. We now show that (23) holds for $\theta = \theta_i$ and large enough $n$ by induction. Assume it holds for $\theta_i$ and let $\theta' = \alpha^{-1}\theta_i + (1-\alpha^{-1})(\alpha-2) \in (\theta_{i+1}, \alpha-2)$. Choose $N(\omega)$ ($<\infty$ a.s.) so that for $n \ge N$ (23) holds for $\theta = \theta_i$, (20) holds and (19) holds for $\theta = \theta'$ and $q = \alpha - 1 - \theta'$. If $n \ge N$, $t \in S^+(e^{-n}n^{-1},1)$ and $a \in \mathbb{R}$, choose $b \in S(e^{-n^q})$ such that $|b-a| \le e^{-n^q} \le e^{-n}$. Then

$$L^a_{t+e^{-n}} - L^a_t$$

$$\le |L^b_{t+e^{-n}} - L^b_t| + c_8((\sup_x L^x_{t+e^{-n}} - L^x_t) + e^{-n})^{\frac{1}{2}}(n^q)^{1-(\alpha/2)} \quad \text{(by (20))}$$

$$\le n^{-\theta'} + c_8(n^{-\theta_i/2} + e^{-n/2})n^{q(1-(\alpha/2))} \le n^{-\theta_{i+1}}$$

for large enough n. Therefore (23) holds for each $\theta_i$ for large enough n (depending on $\theta_i$ and $\omega$). As $\theta_i$ clearly increases to $\alpha - 2$, an elementary argument now completes the proof. □

I suspect that for $\theta = \alpha - 2$, the left side of (21) equals $c_7^{-1}$ but in this more delicate case our method only produces the correct power of the logarithm.

The above theorem should be compared with

$$\lim_{s\to 0^+} \sup_{t\le 1} (L^0_{t+s} - L^0_t)(\log 1/s)^{\alpha-1}(\log\log 1/s)^{-1} = \pi^{-2} \quad \text{a.s.},$$

a fact that follows easily from Theorem 1 of [5] and easy computations.

Acknowledgement. I wish to thank Martin Barlow and S. James Taylor for a number of enjoyable discussions concerning the local time of Lévy processes.

## References

1. M.T. BARLOW. Continuity of local times for Lévy processes. To appear in Z. *Wahrscheinlichkeitstheorie verw. Gebiete.*

2. M.T. BARLOW, E.A. PERKINS, S.J. TAYLOR. Two uniform intrinsic constructions for the local time of a class of Lévy processes. To appear in *Ill. J. Math.*

3. E.S. BOYLAN. Local times for a class of Markov processes. *Ill. J. Math. 8*, (1964), 19-39.

4. B.E. FRISTEDT, W.E. PRUITT. Lower functions for increasing random walks and subordinators. *Z. Wahrscheinlichkeitstheorie verw. Gebiete 18* (1971), 167-182.

5. B.E. FRISTEDT, W.E. PRUITT. Uniform lower functions for subordinators. *Z. Wahrscheinlichkeitstheorie verw. Gebiete 24* (1972), 63-70.

6. A.M. GARSIA. Continuity properties of multi-dimensional Gaussian processes. *6th Berkeley Symposium on Math. Stat. Prob.*, Vol. 2, 369-374. Berkeley (1970).

7. A.M. GARSIA, E. RODEMICH, H. RUMSEY. A real variable lemma and the continuity of paths of some Gaussian processes. *Indiana Univ. Math. J. 20*, (1970), 565-578.

8. R.K. GETOOR, H. KESTEN. Continuity of local times for Markov processes. *Comp. Math. 24* (1972), 277-303.

9. J. HAWKES. A lower Lipschitz condition for the stable subordinator. *Z. Wahrscheinlichkeitstheorie verw. Gebiete 17* (1971), 23-32.

10. S. OREY, S.J. TAYLOR. How often on a Brownian path does the law of the iterated logarithm fail? *Proc. London Math. Soc. 28, Ser. 3* (1974), 174-192.

11. E.A. PERKINS. The exact Hausdorff measure of the level sets of Brownian motion. *Z. Wahrscheinlichkeitstheorie verw. Gebiete 58* (1981), 373-388.

12. R.M. BLUMENTHAL, R.K. GETOOR. *Markov Processes and Potential Theory*. Academic Press, New York, 1968.

EDWIN PERKINS
Department of Mathematics
University of British Columbia
Vancouver, B.C.
Canada V6T 1Y4

*Seminar on Stochastic Processes, 1984*
Birkhäuser, Boston, 1986

# CONVERGENCE IN ENERGY AND THE SECTOR CONDITION FOR MARKOV PROCESSES

by

Z. R. POP-STOJANOVIC

## Introduction

In an earlier paper [2], p. 148, dealing with properties of the energy of Markov processes, M. Rao and the author have shown that the so-called sector condition introduced by M. L. Silverstein [4], p. 17, is sufficient for the regularity of the limit potential. This paper will explain the meaning of the sector condition in relation to convergence in energy for certain classes of Markov processes.

Notations used here are those of Blumenthal-Getoor [1]. Thus $X = (\Omega, F, F_t, X_t, \theta_t, P^x)$ denotes a transient Hunt process on a locally compact second countable state space $(E, E)$ with transition semi-group $(P_t)$. It is assumed that there is an excessive reference measure $\xi$; this measure will be denoted by dx. For non-negative functions f and g, (f,g) will denote $\int f(x)g(x)\,dx$. With $(U^\alpha)_{\alpha>0}$ we denote the resolvent operator of X, i.e. there is a non-negative function $u^\alpha(x,y) \in E^* \times E^*$ such that

(i) $x \to u^\alpha(x,y)$ is $\alpha$-excessive for each y;

(ii) $y \to u^\alpha(x,y)$ is $\alpha$-coexcessive for each x;

(iii) $U^\alpha f(x) = \int u^\alpha(x,y)\, f(y)\, dy$, $\quad f \in bE$.

In addition, the family of co-resolvent operators $(\hat{U}^{\alpha})_{\alpha>0}$ is defined by

$$\text{(iv)}\quad f\hat{U}^{\alpha}(y) = \hat{U}^{\alpha}f(y) = \int u^{\alpha}(x,y)f(x)\,dx\ , \quad f \in b\mathcal{E}\ .$$

However, we do not assume in this paper the existence of the dual process $\hat{X}$ . The density for the potentials $U \equiv U^0$, $\hat{U} \equiv \hat{U}^0$ will be denoted by $u(x,y)$ when it exists. Finally, if $\nu$ is a measure on $(E,\mathcal{E})$ and $f \in b\mathcal{E}$, then $(f,\nu) \equiv (\nu,f) = \int f\,d\nu$ . Also, it is asssumed that $U$ is proper, so that every excessive function is the increasing limit of a sequence $(Uf_n)_n$ .

DEFINITION. *Let* $\alpha \geq 0$ . *If* $(U^{\alpha}|f|,|f|) < +\infty$ , *one says that* $U^{\alpha}f$ *has finite energy. Then,*

$$(1)\qquad ||Uf||_{\alpha,e} = (U^{\alpha}f,f)^{\frac{1}{2}} \geq 0$$

*is called the* $\alpha$*-energy of* $U^{\alpha}f$; $||Uf||_{0,e} = ||Uf||_e$ .

## Sector Condition

M. L. Silverstein [4] introduced the following so-called "sector condition": let $Uf$ and $Ug$ be any two admissible functions. Then, we say that the *sector condition* holds if there exists a constant $M > 0$ such that

$$\text{(SC)}\qquad |(Uf,g)| \leq M\,||Uf||_e||Ug||_e$$

In [2] it is shown that (SC) is sufficient for the regularity of natural potentials. The main objective of this paper is to give a new characterization of the sector condition which will help in understanding its meaning. Toward this goal let us consider the operator

$$(2)\qquad V = \int_0^{\infty} \frac{1}{\sqrt{\pi t}}\,P_t\,dt\ ,$$

where $(P_t)$ denotes the transition semi-group of the process X . Then, V is the *potential* operator of the semi-group obtained by subordination of $(P_t)$ using the one-sided stable semi-group of order 1/2 .

LEMMA. *For any admissible* $f \geq 0$ ,

$$V^2 f = Uf \; . \tag{3}$$

PROOF. Recall that

$$Uf(x) = \int_0^\infty P_t f(x)\,dt, \quad P_t f(x) = \int_E P_t(x,dy) f(y)$$

Then

$$V^2 f(x) = \int_0^\infty \frac{1}{\sqrt{\pi t}} \; P_t V f(x)\; dt$$

$$= \int_0^\infty \frac{1}{\sqrt{\pi t}} \int_E P_t(x,dy) \int_0^\infty \frac{1}{\sqrt{\pi s}} \; P_s f(y)\; ds\; dt$$

$$= \frac{1}{\pi} \int_0^\infty \int_0^\infty \frac{1}{\sqrt{ts}} \; P_{t+s} f(x)\; ds\; dt \; .$$

After the substitution $t = u$ , $s = v - u$ , where $0 \leq v < +\infty$ , $0 \leq u \leq v$ , the last integral becomes :

$$V^2 f(x) = \frac{1}{\pi} \int_0^\infty P_v f(x)\; dv \int_0^v u^{-1/2}\,(v-u)^{-1/2}\; du = \int_0^\infty P_v f(x)\; dv$$

since, for every $v > 0$,

$$\int_0^v u^{-1/2}(v-u)^{-1/2} du = \pi \; .$$

Thus, $V^2 f = Uf$ as claimed.

REMARK. In terms of resolvent densities, the operator V has the density v given by

(4) $$v(x,y) = \frac{1}{\pi} \int_0^\infty \alpha^{-\frac{1}{2}} u^\alpha(x,y)\, d\alpha \,.$$

Let $\hat{V}$ denote the adjoint operator of V. Then, the density $\hat{v}$ of $\hat{V}$ is given by

$$\hat{v}(x,y) = v(y,x) \,.$$

The following example will motivate the introduction of the operator V. Consider a Lévy process with exponent $\psi$ . For an admissible function f, one has the following formula for the energy of Uf :

(5) $$||Uf||_e^2 = \int \frac{1}{|\psi(\alpha)|} |\hat{f}(\alpha)|^2 \, d\alpha \,,$$

where $\hat{f}$ denotes the Fourier transform of f. However,

(6) $$||Vf||_{L^2}^2 = K \, ||Uf||_e^2 \,,$$

where $K > 0$ is a constant.

Having in mind formulae (5) and (6), one has the following characterization of the (SC) for normal operators.

THEOREM. *Let* U *be a potential operator and assume that* U *is normal. Then,* (SC) *holds if and only if there exists a constant* $M > 0$ *such that*

(7) $$||Uf||_e \le ||Vf||_{L^2} \le M \, ||Uf||_e$$

*for all admissible* f.

PROOF. First, let us observe that if U is a normal operator then V is also a normal operator. In addition,

$$\|Vf\|_{L^2} = \|\hat{V}f\|_{L^2} .$$

Now, using the lemma one gets:

$$(Uf,f) = (V^2f,f) = (Vf,\hat{V}f) \leq \|Vf\|_{L^2}\|\hat{V}f\|_{L^2} = \|Vf\|^2_{L^2} ,$$

or

(8) $$\|Uf\|_e \leq \|Vf\|_{L^2} .$$

Suppose now that $Vf \in L^2$ . Observe that the set

$$\{\hat{V}g \ ; \ g \in L^2\}$$

is dense in $L^2$ since the range of the adjoint operator $\hat{V}$ is in the domain of the infinitesimal generator of the subordinated semi-group.

Now, back to the proof of the theorem. Assume the (SC). Then one has the following inequality : for any admissible functions f and g ,

(9) $$|(Vf,\hat{V}g)| = |(V^2f,g)| = |(Uf,g)| \leq M \|Uf\|_e \|Ug\|_e .$$

Using (8), one gets

$$\|Ug\|_e \leq \|Vg\|_{L^2} = \|\hat{V}g\|_{L^2} .$$

Therefore, (9) becomes

$$|(Vf,\hat{V}g)| \leq M \|Uf\|_e\|\hat{V}g\|_{L^2} ,$$

or

(10) $$(Vf,\hat{V}g/\|\hat{V}g\|_{L^2}) \leq M \|Uf\|_e .$$

By keeping f fixed and varying g one concludes from (10), based on the fact that $\{\hat{V}g\}$ is dense in $L^2$ , that

(11) $$\|Vf\|_{L^2} \leq M \|Uf\|_e .$$

Inequalities (8) and (11) give (7) .

To prove the converse, assume that (7) holds. Let f,g be any two admissible functions. Then

$$|(Uf,g)| = |(V^2f,g)| = |(Vf,\hat{V}g)|$$

$$\leq \|Vf\|_{L^2}\|\hat{V}g\|_{L^2} = \|Vf\|_{L^2}\|Vg\|_{L^2} \leq K\ \|Uf\|_e\|Ug\|_e$$

for some $K > 0$ , that is, the (SC) holds . □

REMARK. This theorem covers all Lévy processes. It can be restated as follows: On the linear space of all functions Uf of finite energy, the energy norm of Uf and $L^2$ norm of Vf *are equivalent if and only if the* (SC) *holds.*

COROLLARY. *For a Lévy process with exponent* $\psi$, *the* (SC) *is equivalent to having*

$$A(1 + \mathrm{Re}(\psi)) \geq |\mathrm{Im}(\psi)| \tag{12}$$

*for some constant* $A > 0$ .

REMARK. Let $\phi_t(z)$ denote the characteristic function of the process, i.e., $\phi_t(z) = E(\exp(i(z,X_t)))$ . Then, it is well-known [1] that

$$\phi_t(z) = \exp(-t\psi(z))$$

and

$$\int_0^\infty e^{-\lambda t}\ \phi_t(z)\ dt = 1/(\lambda + \psi(z))\ ,\quad \lambda > 0\ .$$

PROOF of the corollary. It follows that

$$\|Vf\|^2_{L^2} = \int \frac{|\hat{f}(z)|^2}{|1 + \psi(z)|} \, dz \ .$$

Now if the (SC) holds, one has:

$$\|Vf\|^2_{L^2} \leq M \ (U^1 f, f) \text{ for some } M > 0 \ ,$$

or,

$$\int \frac{|\hat{f}(z)|^2}{|1 + \psi(z)|} \, dz \ \leq \ M \int \frac{[1 + \text{Re}(\psi(z))] \ |\hat{f}(z)|^2}{|1 + \psi(z)|^2} \, dz \ .$$

Since this inequality holds for all admissible f, it follows that

$$M \frac{1 + \text{Re}(\psi(z))}{|1 + \psi(z)|} \geq 1 \ ,$$

or,

$$A(1 + \text{Re}(\psi)) \geq |\text{Im}(\psi)|$$

for some constant $A > 0$ , as claimed. □

The following proposition shows that the (SC) is in a sense "built into" the operator V.

PROPOSITION. *Let* U *be a normal operator. Suppose* $Uf_n \to s$ *almost everywhere as* $n \to +\infty$ *where* s *is a natural potential, and assume that* $(Vf_n)_n$ *is a Cauchy sequence in* $L^2$. *Then* s *is a* regular *potential.*

PROOF. It follows immediately from (8) that $(Uf_n)_n$ is a Cauchy sequence in the energy norm. It has been shown in [2], p. 148, that the only thing one needs to show is

$$\lim_n \ (Uf_n, f_n) = \lim_{m,n} \ (Uf_n, f_m) \ . \tag{14}$$

However, (14) follows from the inequality

$$|(U(f_n - f_m), f_n)| \leq \|\nabla f_n - \nabla f_m\|_{L^2} \|\nabla f_n\|_{L^2} . \qquad \square$$

ACKNOWLEDGMENT. The author wishes to express his profound gratitude to Professor Murali Rao for his valuable suggestions which vastly improved this paper.

## References

[1] R.M. BLUMENTHAL and R.K. GETOOR. *Markov Processes and Potential Theory*. Academic Press, New York, 1968.

[2] Z.R. POP-STOJANOVIC and MURALI RAO. Some Results on Energy. *Seminar on Stochastic Processes 1981*, pp. 135-150. Birkhäuser, Boston, 1981.

[3] Z.R. POP-STOJANOVIC and MURALI RAO. Remarks on Energy. *Seminar on Stochastic Processes 1982*, pp. 229-235, Birkhäuser, Boston, 1983.

[4] M.L. SILVERSTEIN. The Sector Condition implies that semipolar sets are quasi-polar. *Z. Wahrscheinlichkeitstheorie verw. Gebiete*, *41* (1977), 13-33.

Department of Mathematics
University of Florida
Gainesville, Florida 32611

*Seminar on Stochastic Processes, 1984*
Birkhäuser, Boston, 1986

# AN INCREASING DIFFUSION*

by

THOMAS S. SALISBURY

## §1. Introduction

In [2], E. Çinlar and J. Jacod consider, among other things, the problem of whether every continuous strong Markov process of bounded variation is deterministic (a problem apparently also posed by S. Orey). They show that this question is equivalent to that of whether every strong Markov process satisfying an ODE $X_t' = F(X_t)$ is deterministic. At the time of writing [2], they thought they had a proof that this was indeed the case. They later found an error in this proof, but subsequently established the result in the case that $(X_t)$ is one-dimensional. More formally, they can show the following.

(1.1) THEOREM. *Let* $(X_t)$ *be a real valued (time-homogeneous) continuous Hunt process of bounded variation. Then* $X_t$ *is a.s. a deterministic function of* $X_0$.

We will show that this result is false in dimensions bigger than one. In fact, we will produce a *non*-time-homogeneous real valued continuous Hunt process that is not deterministic. It will arise as a

*Research supported in part by NSF Grant DMS 8201128.

deterministic function of a space-time version of a time changed Brownian motion.

I would like to thank Erhan Çınlar and Jean Jacod for advertising their problem. I would also like to thank Burgess Davis for many helpful conversations and invaluable suggestions.

## §2. The Construction

Let $\Omega = [0,\infty) \times C([0,\infty), \mathbb{R})$. The canonical realization of a continuous space-time stochastic process is

$$(\tau_t, B_t)(s,\omega) = (s+t, \omega(t)).$$

Let $\mathcal{F}_t = \sigma(\tau_0, B_s;\ s \le t)$, and set $\mathcal{F} = \mathcal{F}_\infty$. Let $P^{s,x}$ be the law on $(\Omega, \mathcal{F})$ of space-time Brownian motion started at (s,x). As usual, "a.s." means $P^{s,x}$-a.s., for every s, x.

A set of the form

$$[u,v] \times [a,b] \subset [0,\infty) \times \mathbb{R}$$

will be called a *box*. For each union G of finitely many disjoint boxes $[u_i,v_i] \times [a_i,b_i]$, we will define a process $M_t^G$ as follows: It behaves like a Brownian motion till the first hit of G. If this occurs in the i'th box, we 'hold' $M_t^G$ until $\tau_t = v_i$, and then resume Brownian behaviour until the next hit of G, etc. (see Figure (2.4)).

More formally, let $S_0^G = A^G(0) = 0$, and define $S_n = S_n^G$ and $A(t) = A^G(t)$ inductively as follows (we write $S(n) = S_n$, $B(t) = B_t$ interchangeably):

$$S_{n+1} = S_n + \inf\{t > 0:\ (\tau_t + A(S_n), B(S_n + t)) \in G\}$$

$$A(t) = A(S_n) + t - S_n \quad \text{for} \quad t \in (S_n, S_{n+1})$$

$$A(t) = A(t-) + v_i - u_i$$

$$\text{if} \quad t = S_{n+1} \quad \text{and} \quad (\tau_0 + A(t-), B(t)) \in [u_i, v_i] \times [a_i, b_i].$$

Let $\hat{A}^G(t)$ be the (continuous) inverse of $A^G(t)$, and define

$$M^G(t) = B(\hat{A}^G(t)).$$

We will let $G$ become dense in an appropriate manner, and show that these processes converge.

Let

$$\Gamma(t,y,\rho) = \sup\{m;\ \text{there are disjoint open subintervals } I(1)\ldots I(m) \text{ of } [0,t], \text{ each of length } \geq \rho, \text{ such that for each } i \text{ there is an } s \in I(i) \text{ with } |B_s - y| \leq \rho\}$$

$\Lambda(t,y,\zeta,\rho)$ = the number of upcrossings of $[y+\rho, y+\zeta]$ and downcrossings of $[y-\zeta, y-\rho]$ by $(B_s)_{s \in [0,t]}$.

(2.1) LEMMA

(a) For every $t \geq 0$, $E^{s,x}[\rho\Gamma(t,y,\rho)] \to 0$ as $\rho \downarrow 0$, uniformly in $s$, $x$, and $y$.

(b) For every $t \geq 0$ and $\zeta > 0$, $E^{s,x}[\Lambda(t,y,\zeta,\rho)]$ remains bounded uniformly in $s$, $x$, and $y$, as $\rho \downarrow 0$.

PROOF: (b) follows from Doob's up and downcrossing bounds, via the translational invariance of Brownian motion, and the strong Markov property at $\inf\{s;\ B_s = y \pm \rho\}$. Similarly, by the strong Markov property at $\inf\{s;\ B_s = y \pm \rho\}$, we have that

$$E^{s,x}[\rho\Gamma(t,y,\rho)]$$
$$\leq \sup_{|z| \leq \rho} E^{0,z}[2\rho + \inf\{s;\ B_s = \pm\rho\}] + E^{0,0}[\rho\Gamma(t,\rho,\rho)]$$

$$\leq 2\rho + \rho^2 + E^{0,0}[\text{Lebesgue measure of a } \rho\text{-neighborhood of } \{s \leq t;\ B_s \in [0,2\rho]\}].$$

As $\rho \downarrow 0$, the integrand decreases boundedly to the Lebesgue measure of $\{s \leq t;\ B_s = 0\}$, which is zero. □

If $\phi$ is nondecreasing and right continuous (for our purposes, actually continuous), denote its right continuous inverse by $\hat{\phi}$ (that is, $\hat{\phi}(t) = \inf\{s \geq 0;\ \phi(s) > t\}$). If $G$ is the union of finitely many disjoint boxes, we let

$$A^G_\phi(t)(s,\omega) = A^G(t)(s,\omega \circ \phi),$$

and define $\hat{A}^G_\phi$ and $S^G_\phi$ similarly.

Let $\Psi$ consist of all nondecreasing, continuous $\phi$ such that $\phi(0) = 0$ and $\hat{\phi}(t) - t$ is nondecreasing.

(2.2) LEMMA. For $\phi \in \Psi$, and $G$ as above,

$$\hat{A}^G(t) \geq \phi(\hat{A}^G_\phi(t)) \quad \text{for every } t \geq 0.$$

PROOF: Write $A(t)$ and $A'(t)$ for $A^G(t)$ and $A^G_\phi(\hat{\phi}(t))$ respectively. Since $\phi$ is continuous, the inverse of $A'$ is $\phi \circ \hat{A}^G_\phi$, so that it will suffice to show that $A(t) \leq A'(t)$ for every $t$. Let $t(n) = S^G(n)$ for $n \geq 0$. We will show by induction that $A(t) \leq A'(t)$ for every $t \in [t(n), t(n+1))$.

Note first that on any interval $(s,r)$ not containing any time $t(n)$, we have that

$$\begin{aligned} A'(t) - A(t) &\geq (A'(s) + \hat{\phi}(t) - \hat{\phi}(s)) - (A(s) + t - s) \\ &\geq (A'(s) + t - s) - (A(s) + t - s) = A'(s) - A(s). \end{aligned}$$

This starts the induction off, as $\phi(0) = 0$ implies that $A(0) = A_\phi^G(0) \le A'(0)$. Similarly, assuming that $A'(t) \ge A(t)$ on $[t(n),t(n+1))$, we need only show that $A'(t(n+1)) \ge A(t(n+1))$. If $A'(t(n+1)-) \ge A(t(n+1))$, there is nothing to show. Thus assume that $A'(t(n+1)-) < A(t(n+1))$, and let $[u,v] \times I$ be the box of $G$ to which $(\tau_0 + A(t(n+1)-), B_{t(n+1)})$ belongs. Then

$$u \le \tau_0 + A(t(n+1)-) \le \tau_0 + A'(t(n+1)-) \quad \text{(by induction)}$$
$$< \tau_0 + A(t(n+1)) = v.$$

But in this case,

$$\tau_0 + A'(t(n+1)) \ge \tau_0 + A_\phi^G(\hat{\phi}(t(n+1)-)) = v = \tau_0 + A(t(n+1)),$$

showing the result. □

Now let $\Phi(\delta)$ consist of all nondecreasing continuous functions $\phi: [0,\infty) \to [0,\infty)$ such that $|\phi(t) - t| < \delta$ for every $t \ge 0$. In the following lemma, it is more or less clear that some $\delta$ will work. Specifying that $\delta$ and then verifying the result is a tedious task however, and the proof has been relegated to the next section; readers are advised to omit it!

(2.3) LEMMA. For $G$ as above, and $\varepsilon > 0$

$$P^{s,x}(\sup\{|\hat{A}^G(t) - \hat{A}_\phi^G(t)|;\ t \ge 0,\ \phi \in \Phi(\delta)\} \ge \varepsilon) \to 0$$

as $\delta \downarrow 0$, uniformly in $s, x$.

PROOF: See §3.

Now let $\Sigma\varepsilon(n) < \infty$. Set $G(0) = \phi$, $\xi(0) = 1$, and define $G(n)$ and

$H(n)$ inductively as follows:

Use Lemma (2.3) to find $\xi(n) \in (0,\xi(n-1)/2)$ so that

$$\sup_{s,x} P^{s,x}(\sup\{|\hat{A}^{G(n-1)}(t) - \hat{A}_{\phi}^{G(n-1)}(t)|;$$

$$t \geq 0,\ \phi \in \Phi(4\xi(n))\} \geq \varepsilon(n)) < \varepsilon(n).$$

Then use that lemma again, to find $\delta(n) \in (0,\xi(n))$ so that

$$\sup_{s,x} P^{s,x}(\sup\{|\hat{A}^{G(n-1)}(t) - \hat{A}_{\phi}^{G(n-1)}(t)|;$$

$$t \geq 0,\ \phi \in \Phi(\delta(n))\} \geq \xi(n)) < \varepsilon(n).$$

Now use Lemma (2.1) to find $\rho(n) \in (0,2^{-n})$ so that

$$\sup_{s,x,y} P^{s,x}\big(3\rho(n)[\Gamma(n,y,\rho(n)) + \Lambda(n,y,2^{-n},\rho(n))] + 1]$$

$$\geq \frac{\delta(n)}{2^{2n+1}}\big) < \frac{\varepsilon(n)}{2^{2n+1}}.$$

Let $H(n)$ be the union of those boxes

$$([\rho(n)(4k+(-1)^j+1),\ \rho(n)(4k+(-1)^j+4)] \cap [0,n])$$

$$\times\Big[\frac{2j+1}{2^n} - \rho(n),\ \frac{2j+1}{2^n} + \rho(n)\Big]$$

not intersecting $G(n-1)$, for which $k \geq 0$ and $|j| < 2^{2n}$. Let $G(n) = G(n-1) \cup H(n)$.

Recall (cf. [1]) that a continuous $\mathbb{R}^2$-valued strong Markov process is a *Hunt process* if its semigroup maps Borel functions to Borel functions.

(2.4)

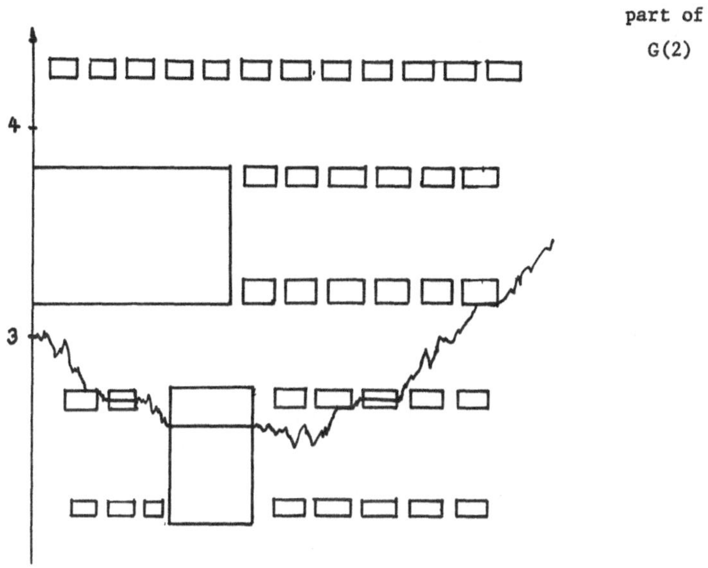


(2.5) THEOREM.

(a) $\hat{A}^{G(n)}$ *decreases to a function* $\hat{A}$ *as* $n \to \infty$. *Moreover, the convergence is a.s. uniform.*

(b) $\hat{A}(t)$ *is continuous in* $t$ *and* $\to \infty$ *as* $t \to \infty$, *a.s.*

(c) $\hat{A}(r)$ *is an* $(\mathcal{F}_{t+})$ *stopping time for each* $r \geq 0$. *Let* $\mathcal{G}_t = \mathcal{F}_{\hat{A}(t)+}$. *If* $T$ *is a* $(\mathcal{G}_{t+})$ *stopping time then* $\hat{A}(T)$ *is an* $(\mathcal{F}_{t+})$ *stopping time, and* $\mathcal{G}_{T+} \subset \mathcal{F}_{\hat{A}(T)+}$.

(d) *Let* $M_t = B_{\hat{A}(t)}$. *Then for all* $(\mathcal{G}_{t+})$ *stopping times* $T$ *we have that* $M_{T+t} = M_t(\tau_T, B_{\hat{A}(T)+}.)$ *for every* $t$, *a.s.*

(e) $((\tau_t, M_t), \mathcal{G}_{t+}, P^{s,x})$ *is a continuous Hunt process.*

[*Note*: the monotone convergence of (a) is not strong enough. It will be crucial that $\hat{A}$ is not constant, and for that we need uniform convergence.]

PROOF: Let $\nu(n)$ be the total time that $(\tau_t, M_t^{G(n)})$ spends in

H(n). Fix $n \geq 1$ for now, and let

$$\hat{\phi}(t) = t + \sum_{r\leq t} (A^{G(n)}(r) - A^{G(n)}(r-))1_{\{(\tau_0 + A^{G(n)}(r-), B_r) \in H(n)\}}.$$

Let $\phi$ be the (continuous) inverse of $\hat{\phi}$. Then $\phi \in \Psi$ and

$$A^{G(n)}(t) = A_\phi^{G(n-1)}(\hat{\phi}(t)),$$

so that also

$$\hat{A}^{G(n)}(t) = \phi(\hat{A}_\phi^{G(n-1)}(t)).$$

The monotone convergence part of (a) now follows by Lemma (2.2).

Now suppose that $[u,v] \times [y-\rho(n), y+\rho(n)]$ is a box of H(n) of length $3\rho(n)$ (this is a restriction only if $v = n$), which is hit by $(\tau_t, M_t^{G(n)})$. Let

$$I = (\hat{A}^{G(n)}(u), \hat{A}^{G(n)}(v) + \rho(n)).$$

If $B_s \in (y-2^{-n}, y+2^{-n})$ for $s \in I$, then $|I| > \rho(n)$. If not, then $(B_s)_{s\in I}$ makes an upcrossing of $[y+\rho(n), y+2^{-n}]$, or a downcrossing of $[y-2^{-n}, y-\rho(n)]$. Thus

$$\#\{\text{boxes of } H(n) \text{ hit by } (\tau_t, M_t^{G(n)})\}$$

$$\leq \sum_{|j|<2^{2n}} [N(n, \frac{2j+1}{2^n}, \rho(n)) + L(n, \frac{2j+1}{2^n}, \frac{1}{2^n}, \rho(n)) + 1],$$

and hence

$$P^{s,x}(\nu(n) \geq \delta(n)) \leq \sum_{|j|<2^{2n}} \frac{\varepsilon(n)}{2^{2n+1}} < \varepsilon(n).$$

If $\nu(n) < \delta(n)$ then $\phi \in \Phi(\delta(n))$. Therefore

$$(2.6)\qquad P^{s,x}(\sup_t|\hat{A}^{G(n-1)}(t) - \hat{A}^{G(n)}(t)| \geq 2\xi(n))$$

$$\leq P^{s,x}(\sup_t|\hat{A}^{G(n-1)}(t) - \hat{A}_\phi^{G(n-1)}(t)| \geq \xi(n)) < 2\varepsilon(n),$$

so that the remainder of (a) follows from the easy half of the Borel-Cantelli Lemma. Part (b) follows in turn, as each $\hat{A}^{G(n)}$ is eventually linear with unit slope.

To show (c), let $A$ be the right continuous inverse of $\hat{A}$. Then $A(t-) = \lim A^{G(n)}(t-)$ for every $t$. By construction, each $A^{G(n)}(t)$ is adapted to $(\mathcal{F}_t)$, hence so is $A(t-)$. Thus $A(t)$ is adapted to $(\mathcal{F}_{t+})$, and

$$\{\hat{A}(r) \leq t\} = \{r \leq A(t)\} \in \mathcal{F}_{t+},$$

so that $\hat{A}(r)$ is an $(\mathcal{F}_{t+})$ stopping time. We may therefore define $\mathcal{G}_t = \mathcal{F}_{\hat{A}(t)+}$. Then also

$$\{A(t) < s\} = \{t < \hat{A}(s)\} \in \mathcal{F}_{\hat{A}(s)+} = \mathcal{G}_s,$$

so that $A(t)$ is a $(\mathcal{G}_{s+})$-stopping time.

Let $T$ be a $(\mathcal{G}_{s+})$-stopping time. For $B \in \mathcal{G}_{T+}$, we have that

$$B \cap \{\hat{A}(T) \leq t < \hat{A}(s)\} = B \cap \{T \leq A(t) < s\} \in \mathcal{G}_s = \mathcal{F}_{\hat{A}(s)+},$$

so that for each $r$ also

$$B \cap \{\hat{A}(t) \leq t < \hat{A}(s) \leq r\} \in \mathcal{F}_{r+}.$$

Taking the union over $s \in \mathcal{G}$, we see that for each $r > t$,

$$B \cap \{\hat{A}(T) \leq t\} \in \mathcal{F}_{r+}.$$

Thus $\hat{A}(T)$ is an $(\mathcal{F}_{t+})$ stopping time, and $\mathcal{G}_{T+} \subset \mathcal{F}_{\hat{A}(T)+}$.

Turning to (d), let T be a $(\mathcal{G}_{t+})$ stopping time. Then $\tau_T \in \mathcal{F}_{\hat{A}(T)+}$, so that by the strong Markov property of $(B_t)$ at $\hat{A}(T)$,

$$P^{s,x}(\sup\{|\hat{A}^{G(n)}(t)(\tau_T, B_{\hat{A}(T)+.}) - \hat{A}_\phi^{G(n)}(t)(\tau_T, B_{\hat{A}(T)+.})|;$$

$$t \geq 0,\ \phi \in \Phi(4\xi(n+1))\} \geq \varepsilon(n+1))$$

$$= E^{s,x}[P^{\tau_T, B_{\hat{A}(T)}}(\sup\{|\hat{A}^{G(n)}(t) - \hat{A}_\phi^{G(n)}(t)|;\ t \geq 0,$$

$$\phi \in \Phi(4\xi(n+1))\} \geq \varepsilon(n+1))] < \varepsilon(n+1).$$

Thus by the Borel-Cantelli Lemma and (2.6), we have that a.s. there is an $n_0$ such that for $n \geq n_0$,

$$|\hat{A}^{G(n)}(t) - \hat{A}^{G(n+1)}(t)| < 2\xi(n+1) \quad \text{for every } t, \text{ and}$$

$$\sup\{|\hat{A}^{G(n)}(t)(\tau_T, B_{\hat{A}(T)+.}) - \hat{A}_\phi^{G(n)}(t)(\tau_T, B_{\hat{A}(T)+.})|;$$

$$t \geq 0,\ \phi \in \Phi(4\xi(n+1))\} < \varepsilon(n+1).$$

Let $n \geq n_0$. Then

$$|\hat{A}^{G(n)}(T) - \hat{A}(T)| \leq \sum_{k\geq n} |\hat{A}^{G(k)}(T) - \hat{A}^{G(k+1)}(T)|$$

$$< 2\sum_{k\geq n} \xi(k+1) < 2\sum_{k\geq n} 2^{n-k}\xi(n+1) = 4\xi(n+1).$$

By (a) we have that $\hat{A}(T) \leq \hat{A}^{G(n)}(T)$, so

$$\phi(t) = t + \hat{A}^{G(n)}(T) - \hat{A}(T) \in \Phi(4\xi(n+1)).$$

We have that $\phi(\hat{A}(T)+t) = \hat{A}^{G(n)}(T) + t$, and by construction,

$$\hat{A}^{G(n)}(t)(\tau_T, B_{\hat{A}^{G(n)}(T)+.}) = \hat{A}^{G(n)}(T+t) - \hat{A}^{G(n)}(T) \quad \text{for every } t.$$

Thus

$$|\hat{A}^{G(n)}(t)(\tau_T, B_{\hat{A}(T)+.}) - \hat{A}^{G(n)}(T+t) + \hat{A}^{G(n)}(T)| < \varepsilon(n+1).$$

Letting $n \to \infty$, we see that

$$\hat{A}(t)(\tau_T, B_{\hat{A}(T)+.}) = \hat{A}(T+t) - \hat{A}(T) \quad \text{for every } t, \text{ a.s.},$$

showing (d).

Finally, let T be a $(\mathcal{G}_{t+})$ stopping time, $Z \in \mathcal{G}_{T+}$ bounded, and f bounded and measurable on path space. Then

$$\begin{aligned} E^{s,x}[Zf((\tau_{T+.}, M_{T+.}))] &= E^{s,x}[Zf((\tau.,M.)(\tau_T, B_{\hat{A}(T)+.}))] \\ &= E^{s,x}[ZE^{\tau_T, M_T}[f((\tau.,M.))]]. \end{aligned}$$

This shows the strong Markov property. Because of their Brownian heritage, the transition function for each of the $(\tau_t, M_t^{G(n)})$ can (given sufficient time) be found more or less explicitly. Thus each of these is a continuous Hunt process, hence so is $(\tau_t, M_t)$. □

We will need the following fact later:

(2.7) LEMMA. Let $(s,y) \in [u,v] \times [a,b] \subset G(n)$. Then

$$P^{s,y}(M_t = y \quad \text{for} \quad t \in [0, v-s]) = 1.$$

PROOF. By construction,

$$\hat{A}^{G(k)}(v-s) = 0, \quad P^{s,y} - \text{a.s.}$$

for each $k \geq n$. Thus also $\hat{A}(v-s) = 0$, $P^{s,y}$ - a.s., showing the result. □

Now enumerate the boxes of $G = \cup G(n)$ as $[t_n - x_n, t_n + x_n] \times [a_n, b_n]$. Choose $y_n > 0$ so that

$$\beta = \sum_{n=1}^{\infty} y_n / x_n < 1.$$

Let

$$f(y) = \begin{cases} 0, & y \le 0 \\ y, & 0 \le y \le 1 \\ 1, & y \ge 1 \end{cases}$$

$$g_m(t,y) = 1 + t - \sum_{n=1}^{m} y_n f\left(\frac{y - a_n}{b_n - a_n}\right)\left(\left[1 - \left|\frac{t - t_n}{x_n}\right|\right] \vee 0\right) .$$

Each $g_m$ is continuous on $[0,\infty) \times \mathbb{R}$. Since $|g_{m+1} - g_m| \le y_m$, and

$$\sum y_m \le \sum \frac{y_m}{x_m} < \infty$$

we see that the $g_m$ converge uniformly to a continuous function $g$.

Fix $s$ for the moment. Each $g_m(s,\cdot)$ is nondecreasing and moreover is strictly increasing on any interval $(a_n, b_n)$ with $n \le m$ and $s \in (t_n - x_n, t_n + x_n)$. By construction, the union of such intervals becomes dense in $\mathbb{R}$ as $m \to \infty$, so that in fact each $g(s,\cdot)$ is strictly increasing, hence one to one. Let $h(s,\cdot)$ be its inverse. Then $h$ is continuous. For $(s,x) \in [0,\infty) \times \mathbb{R}$, define

$$Q^{s,x} = P^{s,h(s,x)} .$$

(2.8) THEOREM. *Let* $X_t = g(\tau_t, M_t)$. *Then* $(X_t)$ *is continuous and nondecreasing a.s., yet*

$$((\tau_t, X_t), \mathcal{G}_{t+}, Q^{s,x})$$

*is a nondeterministic Hunt process.*

PROOF: Since $\hat{A}(t) \le t$ for every $t$, $(M_t)$ is a $Q^{s,x}$-martingale for each $s, x$. It is nonconstant by (b) of Theorem (1.5), hence is nondeterministic. Therefore since each $g(r,\cdot)$ is one to one, $(\tau_t, X_t)$ is both nondeterministic and strong Markov. It is continuous since $g$ and $(\tau_t, M_t)$ are. It is therefore Hunt, since $g$ and $h$ are Borel. Thus all that remains to be shown is that $(X_t)$ is a.s. nondecreasing. It suffices to show that each $g_m(\tau_t, M_t)$ is.

By definition each $g_m(\cdot, y)$ is absolutely continuous with 'derivative'

$$1 - \sum_{n=1}^{m} \frac{y_n}{x_n} f\Big(\frac{y-a_n}{b_n-a_n}\Big) \operatorname{sign}(t_n - \cdot) \ge 1 - \sum_{n=1}^{m} \frac{y_n}{x_n} > 1 - \beta > 0.$$

Thus each $g_m(\cdot, y)$ is increasing. Moreover, $[0,\infty)\times \mathbb{R}$ may be decomposed into finitely many boxes (now allowing infinite sides) which either are subsets of some $G(n)$, or on which $g_m(t,y)$ does not depend on $y$. It is clear that $g_m(\tau_t, M_t)$ increases while $(\tau_t, M_t)$ remains in any rectangle of the latter type, and by Lemma (2.7) it also increases on rectangles of the former type. □

## §3. Proof of Lemma (2.3)

### §3.1 *Definitions and outline of proof:*

Let $m \ge 1$ be the number of boxes of $G$, and write

$$G = \bigcup_{i=1}^{m} [u_i, v_i] \times [a_i, b_i].$$

For convenience, let $I_i = [a_i, b_i]$. Choose $\gamma \in (0,1)$ so that $|a_i - b_j| > \gamma$ whenever $I_i \cap I_j = \phi$, and $|u_i - v_j| > \gamma$ otherwise (thus all boxes have length at least $\gamma$ in the time direction, and are at least distance $\gamma$ apart).

Given $\lambda \in (0,1)$, choose $\eta \in (0,(\gamma/5) \wedge \varepsilon)$ so that

$$P^{0,0}(|B_t| < \gamma \quad \text{for} \quad t \in [0,2\eta]) > 1 - \frac{\lambda}{4} \quad \text{and}$$

$$P^{0,0}((t,B_t) \text{ hits } D) < \lambda/8m^2, \quad \text{for every set } D$$

of the form $[s,s+4\eta] \times \{y\}$ not intersecting the $\gamma/2$-neighborhood of $(0,0)$.

Then choose $\delta \in (0,\eta/3)$ so that

$$P^{0,0}(B_t = 0 \quad \text{for some} \quad t \in (2\delta,\eta-\delta)) > 1 - \frac{\lambda}{4}\,.$$

Write $A(t) = A^G(t)$. Let

$$T(1) = \inf\{t > 0;\ (\tau_t,B_t) \in \bigcup_i [u_i - \delta,\ v_i + \delta] \times I_i\},$$

$$J(1) = \begin{cases} 0, & \text{if} \quad T(1) = \infty \\ i, & \text{if} \quad (\tau_{T(1)},B_{T(1)}) \in [u_i - \delta,\ v_i + \delta] \times I_i, \end{cases}$$

and define $T(n)$, $J(n)$ inductively for $n \geq 2$ by

$$T(n) = \inf\{t > T(n-1);\ (\tau_0 + A(t-),B_t) \in \bigcup_{i \in \{1\cdots m\}\setminus\{J(1)\cdots J(n-1)\}} [u_i - 2\eta,\ v_i + 2\eta] \times I_i\}$$

$$J(n) = \begin{cases} 0, & \text{if} \quad T(n) = \infty \\ i, & \text{if} \quad (\tau_0 + A(T(n)-),B_{T(n)}) \in [u_i - 2\eta,\ v_i + 2\eta] \times I_i. \end{cases}$$

Let

$$C'(1) = \{T(1) < \infty,\ \tau_{T(1)} \in [u_{J(1)} - \delta,\ v_{J(1)} - \eta],$$
$$(\tau_t,B_t) \text{ hits } (\tau_{T(1)} + 2\delta,\tau_{T(1)} + \eta - \delta) \times I_{J(1)}, \quad \text{and}$$
$$|B_t - B_{T(1)}| < \gamma \quad \text{for} \quad t \in [T(1),\ T(1) + 2\eta]\},$$

$$C''(1) = \{T(1) < \infty,\ \tau_{T(1)} \in (v_{J(1)} - \eta,\ v_{J(1)} + \delta], \text{ and}$$
$$|B_t - B_{T(1)}| < \gamma \text{ for } t \in [T(1), T(1) + 2\eta]\},$$

$$C(1) = \{T(1) = \infty\} \cup C'(1) \cup C''(1).$$

For $n \geq 2$, let

$$C'(n) = \{T(n) < \infty \text{ and } \tau_0 + A(T(n)-) \in [u_{J(n)} + 2\eta,\ v_{J(n)} - 2\eta]\},$$

$$C''(n) = \{T(n) < \infty \text{ and } B_t \in (a_{J(n)}, b_{J(n)}) \text{ for } t \in [T(n), T(n) + 4\eta]\},$$

$$y(n) = \begin{cases} 1 & \text{on } C'(n) \\ 2 & \text{on } C''(n) \\ 0 & \text{elsewhere}, \end{cases}$$

$$C(n) = \{T(n) = \infty\} \cup C'(n) \cup C''(n),$$

$$C = C(1) \cap \cdots \cap C(m).$$

Let $S = S^G$, and

$$L = 1_{C''(1)},\ K = 1_{C''(1) \cap \{(\tau_t, B_t) \text{ hits } [u_{J(1)}, v_{J(1)}] \times I_{J(1)}\}}.$$

A straightforward induction shows that the following conditions hold:

(3.1) If $K = 1$ then $S(1) = T(1)$ and $\tau_0 + A(S(1)) = v_{J(1)} < \tau_0 + A(S(1)-) + \eta$.

(3.2) $S(1) \in [T(1), T(1)+\eta)$ and

$$(\tau_0 + A(S(1)), B_{S(1)}) \in \{v_{J(1)}\} \times I_{J(1)} \text{ on } C'(1).$$

(3.3) $S(K - L + n) = T(n)$ on $C \cap C'(n)$ for $n \geq 2$.

(3.4) $S(K - L + n) = T(n) + 2\eta < T(n+1)$ and

$$(\tau_0 + A(S(K-L+n)-), B_{S(K-L+n)}) \in \{u_{J(n)}\} \times (a_{J(n)}, b_{J(n)})$$

on $C \cap C''(n)$, for $n \geq 2$.

(3.5) $S(K - L + n) = \infty$ on $C \cap \{T(n) = \infty\}$, for $n \geq 1$.

We will show from this (see §3.2 below) that $P^{s,x}(C) > 1 - \lambda$ for every $s$ and $x$.

Now let $\phi \in \Phi(\delta)$, and write

$$A' = A^G_\phi,\ S' = S^G_\phi,$$

$$K' = 1_{C''(1) \cap \{(\tau_t, B_\phi(t)) \text{ hits } [u_{J(1)}, v_{J(1)}] \times I_{J(1)}\}}.$$

Conditions analogous to (3.1)-(3.5) hold for these objects as well, but we will state matters slightly differently; we will show that the following conditions hold on $C$ (for $n = 1 \ldots m$):

(3.6) If $K' = 1$ then $T(1) \leq \phi(S'(1)) < T(2)$, $\tau_0 + A'(S'(1)) = v_{J(1)}$,

and $\tau_0 + A'(S'(1)-) \in (\tau_0 + A(T(1)-) - \delta,\ v_{J(1)}]$

(3.7) $T(L+n) \leq \phi(S'(K'+n))$, and if $T(L+n) < \infty$ then also

$\phi(S'(K'+n)) < T(L+n+1)$.

(3.8) If $T(L+n) < \infty$ then

$$(\tau_0 + A'(S'(K'+n)), B_{\phi(S'(K'+n))}) \in \{v_{J(L+n)}\} \times I_{J(L+n)}.$$

(3.9) If $T(L+n) < \infty$ then $|S(K+n) - S'(K'+n)| < \eta$.

The induction step will be shown in §3.3, and the induction started off (and (3.6) shown) in §3.4 and §3.5.

Lastly, we will show in §3.6 that from these conditions it follows that

(3.10) $\sup_t |\hat{A}(t) - \hat{A}'(t)| < \varepsilon$ on $C$,

completing the proof of the lemma.

§3.2 *Proof that* $P^{s,x}(C) > 1 - \lambda$:

Fix s and x. For $i \neq j$, each component of $[u_j - 2\eta, u_j + 2\eta) \times \{a_j, b_j\}$ or $(v_j - 2\eta, v_j + 2\eta] \times \{a_j, b_j\}$ is of distance at least

$$\gamma - 2\eta > \delta + \frac{\gamma}{2}$$

from $\{v_i\} \times I_i$. By (3.1) we have that

$$A(T(1)) \in [v_{J(1)}, v_{J(1)} + \delta] \quad \text{on} \quad C''(1).$$

By (3.2)-(3.5), we may therefore apply the strong Markov property at either $T(1)$ or $S(n)$, and obtain that for $n = 1 \ldots m-1$

$$P^{s,x}(C(n) \setminus C(n+1))$$

$$\leq \sum_{i=1}^{m} P^{s,x}(J(n)=i) \sup_{\substack{r \in [v_i, v_i+\delta) \\ y \in I_i}} P^{r,y}((\tau_t, B_t)$$

$$\text{hits} \bigcup_{j \neq i} ([u_j - 2\eta, u_j + 2\eta) \cup (v_j - 2\eta, v_j + 2\eta]) \times \{a_j, b_j\})$$

$$< 4(m-1) \frac{\lambda}{8m^2} < \frac{\lambda}{2m} .$$

Similarly,

$$P^{s,x}(\Omega \setminus C(1)) \leq \sum_{i=1}^{m} P^{s,x}(J(1)=i) \sup_{\substack{r \in [u_i - \delta, v_i + \delta] \\ y \in I_i}}$$

$$[P^{r,y}(|B_t - y| \geq \gamma \text{ for some } t \in [0, 2\eta]$$

$$+ P^{r,y}(B_t \neq y \text{ for any } t \in (2\delta, \eta - \delta))] < \frac{\lambda}{2} .$$

Thus

$$P^{s,x}(C) > 1 - \frac{(m-1)\lambda}{2m} - \frac{\lambda}{2} > 1 - \lambda.$$

§3.3 *Proof of* (3.7)-(3.9); *induction step*:

Let $k \geq 1$, and suppose that (3.6) holds, as do (3.7)-(3.9) for $n = 1 \ldots k$. Then

$$S'(K'+k+1) = \inf\{t > S'(K'+k);\ (v_{J(L+k)} + t - S'(K'+k),\ B_{\phi(t)}) \in G\}.$$

$\phi$ cannot be constant on $[S'(K'+k),\ S'(K'+k+1)]$ (if it were, then $[\tau_0 + A'(S'(K'+k)), \tau_0 + A'(S'(K'+k)) + S'(K'+k+1) - S'(K'+k)] \times \{B_{\phi(S'(K'+k))}\}$ would stretch from one box of $G$ to another, hence would be of length at least $\gamma$, whereas

$$S'(K'+k+1) - S'(K'+k)$$
$$= S'(K'+k+1) - \phi(S'(K'+k+1)) + \phi(S'(K'+k)) - S'(K'+k) < 2\delta < \gamma).$$

Thus

$$\phi(S'(K'+k+1)) = \inf\{r > \phi(S'(K'+k));\ (v_{J(L+k)} + t - S'(K'+k), B_r) \in G$$
$$\text{for some } t \text{ with } \phi(t) = r\}.$$

By induction, if $\phi(t) = r$, then

$$(3.11) \qquad |(t - S'(K'+k)) - (r - S(K+k))| < \delta + \eta < 2\eta.$$

Since also $T(L+k) \leq \phi(S'(K'+k))$, and $(\tau_0 + A'(t-), B_{\phi(t)}) \notin [u_{J(n)}, v_{J(n)}] \times I_{J(n)}$ for any $n \leq L+k$ and $t > S'(K'+k)$, we have that $T(L+k+1) \leq \phi(S'(K'+k+1))$.

On $C \cap \{Y(L+k+1) = 1\}$ we have by (3.11) that

$$\phi(S'(K'+k+1)) = T(L+k+1), \text{ and}$$
$$\tau_0 + A'(S'(K'+k+1)-) \in [u_{J(L+k+1)}, v_{J(L+k+1)}],$$

showing (3.7) and (3.8). (3.9) follows by (3.3).

On $C \cap \{Y(L+k+1) = 2\}$, we have by definition that

$$B_{\phi(t)} \in (a_{J(L+k+1)}, b_{J(L+k+1)})$$

$$\text{for} \quad t \in [\hat{\phi}(T(L+k+1)-), \hat{\phi}(T(L+k+1)+4\eta)].$$

Also,

$$(3.12) \quad |\tau_0 + A'(\hat{\phi}(T(L+k+1)-)-) - (u_{J(L+k+1)} - 2\eta)|$$

$$= |v_{J(L+k)} + \hat{\phi}(T(L+k+1)-) - S'(K'+k) - (v_{J(L+k)} + T(L+k+1) - S(K+k)|$$

$$< \delta + \eta < 2\eta - 2\delta,$$

so that

$$u_{J(L+k+1)} - 4\eta < \tau_0 + A'(\hat{\phi}(T(L+k+1)-)-) < u_{J(L+k+1)}$$

$$< \tau_0 + A'(\hat{\phi}(T(L+k+1)-)-) + \hat{\phi}(T(L+k+1)+4\eta) - \hat{\phi}(T(L+k+1)-).$$

Since $4\eta < \gamma$, we conclude that

$$(\tau_0 + A'(t-), B_{\phi(t)})_{t \in [\hat{\phi}(T(L+k+1)-), \hat{\phi}(T(L+k+1)+4\eta)]}$$

hits no box of $G$ other than $[u_{J(L+k+1)}, v_{J(L+k+1)}] \times I_{J(L+k+1)}$, and that it hits that box in the set

$$\{u_{J(L+k+1)}\} \times I_{J(L+k+1)}.$$

Condition (3.8) and the remainder of (3.7) now follow.

To show (3.9), observe that

$$(3.13) \quad |S(K+k+1) - S'(K'+k+1)|$$

$$= |(S(K+k) + u_{J(L+k+1)} - v_{J(L+k)}) - (S'(K'+k) + u_{J(L+k+1)} - v_{J(L+k)})|$$

$$< \eta .$$

Thus, the induction step is shown on $C \cap \{T(L+k+1) < \infty\}$. It holds vacuously on the remainder of C.

§3.4 *Proof of* (3.7)-(3.9) *for* $n = 1$, *on* $C \cap C'(1)$:

$S'(1) = \inf\{t \geq 0;\ (\tau_t, B_{\phi(t)}) \in G\}$, so that $\phi(S'(1)) = \inf\{r \geq \phi(0);\ (\tau_t, B_r) \in G$ for some $t$ with $\phi(t) = r\}$.

Since $|\tau_t - \tau_r| < \delta$ whenever $\phi(t) = r$, we have that

$$\phi(S'(1)) \geq T(1) \text{ everywhere on } \{T(1) < \infty\}. \tag{3.14}$$

Moreover, on $C'(1) \cup C''(1)$ we have that $|B_t - B_{T(1)}| < \gamma$ for $t \in [T(1), T(1) + 2\eta]$, and hence that

$$\begin{array}{l} T(2) > T(1) + 2\eta \text{ and } (\tau_t, B_{\phi(t)})_{t \in [0, T(1)+\eta]} \\ \text{hits no box of } G \text{ other than } [u_{J(1)}, v_{J(1)}] \times I_{J(1)}. \end{array} \tag{3.15}$$

On $C'(1) \cap C$, there is by definition an $r \in (T(1) + 2\delta, T(1) + \eta - \delta)$, with $B_r \in I_{J(1)}$. Then $r = \phi(t)$ for some $t \in (T(1) + \delta, T(1) + \eta)$. Since

$$u_{J(1)} \leq \tau_{T(1)} + \delta < \tau_{T(1)} + \eta \leq v_{J(1)},$$

we see that

$$(\tau_t, B_{\phi(t)}) \in [u_{J(1)}, v_{J(1)}] \times I_{J(1)}.$$

Recalling (3.15), we obtain (3.7) and (3.8). (3.9) follows by (3.2), (3.14) and the fact that $\phi(S'(1)) \leq r < T(1) + \eta$.

§3.5 *Proof of* (3.6) *and of* (3.7)-(3.9) *for* $n = 1$ *on* $C \cap C''(1)$:

Fix a point in $C \cap C''(1)$. If $K' = 1$, then $S'(1) \leq v_{J(1)} - \tau_0 < T(1) + \eta$. Thus (3.6) follows by (3.14) and (3.15). Further (breaking things up into the two cases: that $\tau_0 + A'(S'(1)-)$ belongs to

$(\tau_0 + A(T(1)-)-\delta, \tau_0 + A(T(1)-)]$ or $(\tau_0 + A(T(1)-), v_{J(1)}])$, we have by (3.1) that

$$(3.16) \qquad |A(T(1)+\eta) - A'(T(1)+\eta)| < \eta.$$

Likewise, if $K' = 0$, (3.16) still holds and also $T(2) \wedge S'(1) > T(1)+\eta$. Thus (3.16) holds on all of $C \cap C''(1)$, and irrespective of $K'$,

$$S(K'+1) = \inf\{t > T(1)+\eta; (\tau_0 + A'(T(1)+\eta) + t - T(1)-\eta, B_{\phi(t)}) \in G\}.$$

Thus as in §3.3,

$$\phi(S'(K'+1)) = \inf\{r > \phi(T(1)+\eta); (\tau_0 + A'(T(1)+\eta) + t - T(1)-\eta, B_r)$$
$$\in G \text{ for some } t \text{ with } \phi(t) = r\}.$$

We conclude as in §3.3 that $T(2) \le \phi(S'(K'+1))$, now using the inequality

$$|(A'(T(1)+\eta) + t) - (A(T(1)+\eta) + r)| < \eta + \delta < 2\eta,$$

instead of (3.11).

On $C \cap C''(1) \cap C'(2)$ we may now proceed as in §3.3. On $C \cap C''(1) \cap C''(2)$ we may do likewise, the only modifications being that instead of (3.12) we use that

$$|\tau_0 + A'(\hat{\phi}(T(2)-)-) - (u_{J(2)} - 2\eta)|$$
$$= |\tau_0 + A'(T(1)+\eta) + \hat{\phi}(T(2)) - T(1) - \eta$$
$$- (\tau_0 + A(T(1)+\eta) + T(2) - T(1) - \eta)|$$
$$< \eta + \delta < 2\eta - 2\delta,$$

and that instead of (3.14), we have that

$$|S(K+1) - S'(K'+1)|$$

$$= |T(1)+\eta+u_{J(2)} - A(T(1)+\eta) - (T(1)+\eta+u_{J(2)} - A'(T(1)+\eta))|$$

$$< \eta.$$

Thus by induction, we have shown (3.6)-(3.9) on C, for $n = 1...m$.

§3.6 *Proof of* (3.10):

Fix a point of C. Let $t_0 = 0$, and for $n \geq 1$ set $t_n = v_{J(n)} - \tau_0$ if $T(n) < \infty$ (and $t_n = \infty$ otherwise). If $n \geq 0$ is such that $[t_n, t_{n+1})$ is a finite interval, then conditions (3.1), (3.2), (3.4), (3.6) and (3.8) show that there are $t, t' \in [t_n, t_{n+1}]$ such that $\hat{A}$ and $\hat{A}'$ are linear (with unit slope) on $[t_n, t)$ and $[t_n, t')$ respectively, and then are constant on the remainder of $[t_n, t_{n+1})$. (Note that for $n = 0$ in particular, we may have that $t, t' = t_n$ or $t_{n+1}$). Similarly, if $t_n < \infty = t_{n+1}$, then both $\hat{A}$ and $\hat{A}'$ are linear on $[t_n, t_{n+1})$, with unit slope. Thus we will have that $|\hat{A}(r) - \hat{A}'(r)| < \varepsilon$ for every $r$, provided only that this holds for finite $r$ of the form $t_n$. This in turn follows from (3.1), (3.6) and (3.9) (for example, if $n \geq 1$ then

$$|\hat{A}(v_{J(L+n)} - \tau_0) - \hat{A}'(v_{J(L+n)} - \tau_0)| = |S(K+n) - S'(K'+n)| < \eta < \varepsilon$$

showing (3.10), and hence the Lemma. □

## References

1. R.M. BLUMENTHAL and R.K. GETOOR. *Markov Processes and Potential Theory*. Academic Press, New York, 1968.

2. E. ÇINLAR and J.JACOD. Representations of semimartingale Markov processes in terms of Wiener processes and Poisson random measures. *Seminar on Stochastic Processes 1981*, pp. 159-242. Birkhäuser, Boston, 1981.

Thomas S. Salisbury
Department of Mathematics
York University
Downsview, Ontario M3J 1P3
CANADA

*Seminar on Stochastic Processes, 1984*
Birkhäuser, Boston, 1986

# LARGE DEVIATIONS IN ERGODIC THEORY

by

STEVEN OREY

## 0. Introduction.

The classical example of a large deviation result is Cramer's theorem. It tells us, in a contemporary formulation, that if $Y_1, Y_2, \ldots$ is a sequence of independent real valued random variables with identical distribution function F such that

$$f(\theta) = E[\exp\{\theta Y_1\}[ = \int \exp\{\theta y\} F(dy)$$

is finite for all finite $\theta$, and if $Z_n = (Y_1 + Y_2 + \ldots Y_n)/n$ then

$$k(x) = \sup_\theta [\theta x - \log f(\theta)]$$

satisfies

$$(0.1) \qquad \overline{\lim_{n\to\infty}} \frac{1}{n} \log P[Z_n \in A] \leq -\inf\{k(a): a \in A\}, \quad A \text{ closed}$$

and

$$(0.2) \qquad \underline{\lim_{n\to\infty}} \frac{1}{n} \log P[Z_n \in A] \geq -\inf\{k(a): a \in A\}, \quad A \text{ open.}$$

The law of large numbers asserts that $Z_n$ converges to $E[Y_1]$. Cramer's

---

Partially supported by NSF grant MCS 83-01080

theorem is a refinement: it asserts that if $A$ is away from $E[Y_1]$ $P[Z_n \in A]$ goes down at an exponential rate, and provides the precise exponent.

Our primary goal is to find a result which comes as close as possible to solving the proportion:

X:ergodic theorem = Cramer's theorem: law of large numbers.

We proceed to give an overview of the paper. We hope it will motivate and clarify our work below. More precise statements appear in the body of the paper. All of the definitions introduced in this introduction are repeated in the subsequent sections.

In ergodic theory one starts with $S = (\Omega, F, P, T)$, where $(\Omega, F, P)$ is a probability space and $T: \Omega \to \Omega$ such that $P(T^{-1}(A)) = P(A)$ for $A \in F$. We will demand that $\Omega$ is a separable, metrizable, topological space. In that case we call $S$ a *dynamical system*. Let now $Y$ be a random variable on $(\Omega, F)$, taking for simplicity $Y$ to be real valued, and let $Z_n = (Y + Y\circ T + \ldots + Y\circ T^{n-1})/n$. We seek a function $k: R \to [0,\infty]$ such that (0.1) and (0.2) are satisfied. If such a function exists, such a function satisfying the additional requirement of being lower semicontinuous also exists, and then it is unique (section 1) and will be called the *deviation function* for $(Z_n)$ or the deviation function for $Y$, and it may be denoted by $k_Y$. One feature of the ergodic theorem is that it gives a result (namely a.e. convergence) for all $Y$ in a certain class (namely the integrable functions). Similarly we would like a result which gives us the existence of $k_Y$ for a large class of $Y$. Such a result will indeed follow once one obtains a deviation function for a particular random variable taking values in $M$, the set of probability measures on $(\Omega, F)$ ; here $M$ is considered as a convex subset of the space of all measures on $(\Omega, F)$ topologized by weak convergence. The random variable in question is $\delta$, defined by $\delta_\omega$ = (the unit mass on $\omega$). We set

$$L_n(\omega) = \frac{1}{n}(\delta_\omega + \delta_{T\omega} + \dots + \delta_{T^{n-1}\omega}) .$$

We want a function $K: \mathcal{M} \to [0,\infty]$ which is lower semicontinuous and such that (0.1) and (0.2) are satisfied with $L_n$ and $K$ taking the place of $Z_n$ and $k$ respectively. Such a $K$ will be called the deviation function for a system $S$. Deviation functions may assume the value $+\infty$, but in an important sense only the set on which $K$ is finite matters; we call this set the *deviation carrier*. In all the dynamical systems which we deal with, the deviation carrier is a subset of $\mathcal{M}_T$, the class of $Q$ such that $Q(A) = Q(T^{-1}(A))$ for every $A \in \mathcal{F}$.

We have two methods for obtaining the existence of deviation functions. First a nice class of dynamical systems is isolated for which results can be established directly. Second, we introduce a suitable notion of homomorphism, such that if a dynamical system possesses a deviation function then the homomorphism induces one on the image.

The nice dynamical systems are certain stationary stochastic processes (discrete parameter). We also refer to stationary stochastic processes as stationary shifts. For certain shifts our problem has been solved by Donsker and Varadhan [DV, part 4]. These are shifts which are Markovian and satisfy certain independence conditions. Our first project then is to obtain a nice class of shifts (not necessarily Markovian) for which we can obtain deviation functions by direct methods; this is done in Sections 2,3,5. In Section 6 we give some examples, including dynamical systems which can be obtained as homomorphic images of nice shifts. We wish to mention one, Example 6.2.3, because it is familiar and may indicate quickly the drift of the work. Let S =

$(\Omega, F, P, T)$ where $\Omega$ is the unit interval $(0,1)$, $F$ is the Borel sets of $\Omega$, $T\omega$ is the fractional part of $\omega^{-1}$, and $P$ is given by

$$P(A) = (\log 2)^{-1} \int_A (1+x)^{-1}\, dx\,, \quad A \in F\,.$$

It is a classical result that $P$ is preserved by $T$ (see [B]). This system is studied in connection with continued fraction expansions. Letting $a(\omega)$ be the integer part of $\omega^{-1}$, and $a_n(\omega) = a(T^{n-1}\omega)$, $n = 1,2,\ldots$ the numbers $a_1(\omega)$, $a_2(\omega),\ldots$ are just the partial quotients in the continued fraction expansion of $\omega$. The sequence $a_1(\omega)$, $a_2(\omega),\ldots$ is a stationary sequence, and turns out to be a nice shift, i.e. by direct methods one establishes the existence of a deviation function. The system $S$ is a homomorphic image of this shift, so it too will have a deviation function.

Donsker and Varadhan [DV, part 4] work with a class of Markovian shifts, and they find that the deviation function coincides with a certain entropy function $H(Q)$. This function has been studied earlier in other contexts (see [P]), but when the shift is not Markovian ambiguities in the definition appear, and these have to be handled carefully. We discuss this in Section 2.

The upper bound is treated in Section 3; indeed the beautiful theory developed for this purpose in [DV] essentially carries over. (By a cheap trick any shift can be converted into a Markovian shift. This conversion has a price, including the loss of independence properties, but this turns out to do no harm in obtaining upper bound results.)

To obtain lower bounds a natural first step is to appeal to suitable generalisations of the Shannon-McMillan Theorem. (Such an approach is advocated in [BGZ]). One problem is that at best it is only a first step; (as discussed in Section 5, it is useful in dealing with ergodic $Q$, but further work must be done to obtain a lower bound). A more immediate problem is the availability of generalized Shannon-McMillan Theorem. Such an extension is discussed in Section 4.

In the first part of Section 5 we show that if the assumptions under which we proved the Shannon-McMillan Theorem are supplemented by certain independence assumptions a lower bound can be established. In the second part of Section 5 we follow an alternative approach due to Donsker and Varadhan. They use a "conditional Shannon-McMillan Theorem" which requires no supplementary assumptions. To obtain the desired lower bound however, one again must add independence assumptions. The argument is completed much as in [DV], but the Markov property is not involved.

The lower bound result in the first part of Section 5, Theorem 5.6 is proved under Condition 5.2, whilst the corresponding result in the second part of Section 5 assumes Condition 5.8. These conditions do not appear comparable. The approach via Condition 5.8 is shorter, and readers who wish just to follow this argument can skip Section 4 and the first part of Section 5, through the proof of Corollary 5.7 . However the approach given in the first part of Section 5 is methodologically interesting, and perhaps improvements in the method will allow weakening of the conditions. Furthermore, when this approach applies one gains additional information, (e.g. the infimum in the definition of deviation function can be restricted in this case to the class of stationary ergodic measures).

In Section 1 we introduce the appropriate definitions and establish some basic facts. Our approach here profited from the work of [BZ]. Our greatest debt is to the work [DV], especially part 4. Part 4 of [DV] is also reproduced in the survey article [V]. The surveys [V] and [Az] are recommended for a fine overview of the theory of large deviations and its applications.

*Postscript*. After a preliminary version of this manuscript was circulated some useful information and comments were received. I am indebted to Professor Alan Sokal for a number of helpful comments. He pointed out to me that, in the proof of Theorem 4.9, I only require Condition 5.8, and not a much more stringent form I had previously assumed. He also showed me the counter example mentioned in Example 6.1. The significance of this in our context is that it shows that even in very special situations a deviation function may fail to exist.

I also thank Professor Sokal for pointing out the references (K) and ($P_1$), and Professor Perez for supplying me with copies and remarks.

Finally, I was glad to learn of the extremely interesting work of Y. Takahashi [T]. The "entropy" function $q$ introduced there coincides with the function $-K_*$, where $K_*$ is defined in Section 1. In case a deviation function $K$ exists, $K = K_*$. Takahashi obtains elegant formulas for $q$ in certain cases, and discusses a number of intriguing questions.

## 1. Deviation Functions

Let $(Z_n)$ be a sequence of random variables defined on a probability space $(\Omega,\mathcal{F},P)$ and taking values in a metrisable separable topological space $V$. Take the Borel sets of $V$ as the measurable sets of $V$. (The traditional case to be borne in mind is $Z_n = (Y_1 + Y_2 + \ldots + Y_n)/n$, where $Y_1, Y_2, \ldots$ is a sequence of identically distributed independent random variables and $V$ is a convex subset of a linear topological space; for information in this context see [BZ], [Az]). For a Borel subset of $V$ define

$$\lambda_*(A) = \underline{\lim}_{n\to\infty} \frac{1}{n} \log P[Z_n \in A], \qquad \lambda^*(A) = \overline{\lim}_{n\to\infty} \frac{1}{n} \log P[Z_n \in A],$$

Note that $\lambda_*$ and $\lambda^*$ are non-positive functions, increasing in $A$. A function $k: V \to [0,\infty]$ will be called an *upper deviation function* for $(Z_n)$ if for every Borel set $A$

$$-\inf\{k(a) : a \in A^\circ\} \leq \lambda_*(A) \tag{1.1}$$

where $A^\circ$ denotes the interior of $A$. Similarly $k$ is a *lower deviation function* for $(Z_n)$ if for every Borel set $A$

$$\lambda^*(A) \leq -\inf\{k(a) : a \in \bar{A}\} \tag{1.2}$$

where $\bar{A}$ denotes the closure of $A$. Observe that an upper deviation function provides a lower bound, a lower deviation function gives an upper bound. Given any $k: V \to [0,\infty]$ let

$$k^\vee(a) = \min(k(a), \underline{\lim}_{b\to a} k(b))$$

denote the lower regularisation. Note that $k^\vee$ is lower semicontinuous. The function $k$ will be said to have compact level sets if the set $\{x: k(x) \leq b\}$ is compact for every finite $b$.

Our sequence of random variables $(Z_n)$ satisfies the *compactness condition* if there exists a sequence $(C_n)$ of compact subsets of $V$ such that

$$\lim_{n\to\infty} \lambda^*(V\backslash C_n) = -\infty . \tag{1.3}$$

The following elementary proposition collects some basic facts about deviation functions.

1.1. PROPOSITION. (i) *If* $k$ *is a lower* [*upper*] *deviations function and* $k'$ *is a function from* $V$ *into* $[0,\infty]$ *such that* $k'(a) \leq k(a)\,[k'(a) \geq k(a)]$ *for all* $a \in V$ , *then* $k'$ *is a lower* [*upper*] *deviation function.*

(ii) *if* $k$ *is a lower* [*upper*] *deviation function, then* $k^{\vee}$ *is a lower* [*upper*] *deviation function.*

(iii) *Let* $k^*(a) = -\inf\{\lambda_*(A): A \text{ open}, a \in A\}$ . *Then* $k^*$ *is the least upper deviation function; it is lower semicontinuous.*

(iv) *Let* $k_*(a) = -\inf\{\lambda^*(A): A \text{ open}, a \in A\}$ . *Then, for* $\bar{A}$ *compact,* (1.2) *is satisfied with* $k = k_*$ .

(v) *Suppose the compactness condition* (1.3) *holds and* $k$: $V \to [0,\infty]$ *satisfies* (1.2) *for all* $A$ *such that* $\bar{A}$ *is compact. Then* $k$ *is a lower deviation function.*

(vi) *If* $k$ *is a lower semicontinuous lower deviation function then* $k(a) \leq k_*(a)$ *for all* $a \in V$ .

(vii) *If* $k$ *is both an upper and a lower deviation function then* $k^{\vee}$ *is the unique lower semicontiouous function which is both an upper and a lower deviation function, and* $k^{\vee} = k^* = k_*$ . *If the compactness condition* (1.3) *holds, the condition* $k^* = k_*$ *is sufficient to ensure that* $k^*$ *is both an upper and lower deviation function.*

PROOF. Assertions (i)-(iv) follow almost immediately. Now assume the hypotheses of (v). For $A \in \mathcal{B}$ it must be shown that

$$\lambda^*(A) \leq -\inf\{k(a): a \in \bar{A}\} .$$

So assume $\lambda^*(A) > -\infty$; by (1.3) there exist a compact $C \in \mathcal{B}$ such that $\lambda^*(V \setminus C_n) < \lambda^*(A)$ . Then

$$\lambda^*(A) \leq \lambda^*(\bar{A} \cap C_n) = -\inf\{k(a) : a \in \bar{A} \cap C_n\}$$

$$\leq -\inf\{k(a) : a \in \bar{A}\} .$$

For (vi) introduce a metric on $V$ . Let $a \in V$ and let $B_r$ be the open ball with center $a$ , radius $r$ . Let $\lambda_*(r) = \lambda_*(B_r)$ and $\lambda^*(r) = \lambda^*(\bar{B}_r)$ , $r > 0$ . Note that these functions are negative, and increasing in $r$ ; also

$$k_*(a) = -\lim_{r\to 0} \lambda^*(r) \leq -\lim_{r\to 0} \lambda_*(r) = k^*(a) .$$

Let now $k$ be a lower semicontinuous lower deviation function, and set $g(r) = \inf\{k(x) : x \in \bar{B}_r\}$ . By definition $\lambda^*(r) \leq -g(r)$ , $r > 0$. Now let $r \to 0$ and use the lower semicontinuity of $k$ to conclude $k(a) \leq k_*(a)$ .

The first assertion in (vii) follows from (ii),(iii) and (vi). The final assertion follows from (iii),(iv) and (v). □

The notations $k^*$ and $k_*$ will denote the functions defined in (iii) and (iv). If there is a lower semicontinuous $k$ that is both an upper and a lower deviation function it will be called the *deviation function*; according to (vi) such a function is unique.

The set $\{x \in V: \ k_*^{\vee}(x) < \infty\}$ will be called the *deviation carrier*. By Proposition 1.1, if there is a deviation function $k$ , then $k, k_*, k_*^{\vee}$ and $k^*$ all coincide. In that case the infima in (1.1) and (1.2) effectively extend only over the deviation carrier.

Suppose that there is a measurable map $T: \Omega \to \Omega$ and a random variable $Y$ taking values in $V$, a convex subset of a metrisable, convex linear space, and $Z_n = (Y + Y \circ T + \ldots + Y \circ T^{n-1})/n$. In this situation we shall speak about *deviation functions for* $Y$ to mean deviation functions for the sequence $(Z_n)$.

Given two measurable spaces $(\Omega_1, F_1)$ and $(\Omega_2, F_2)$ and a measurable map $\varphi : \Omega_1 \to \Omega_2$ we shall also denote by $\varphi$ the map which carries any probability measure $P$ on $(\Omega_1, F_1)$ into a probability $\varphi(P)$ on $(\Omega_2, F_2)$ defined by $\varphi(P) = P \circ \varphi^{-1}$.

Suppose that for $i = 1,2$, $(\Omega_i, F_i, P_i)$ are probability spaces, $T_i: \Omega_i \to \Omega_i$ is measurable and $Y_i: \Omega_i \to V_i$ is a random variable, with $V_i$ a convex subset of a metrisable, convex linear space, and $Z_n^{(i)} = (Y_i + Y_i \circ T_i + \ldots Y_i \circ T^{n-1})/n$, $n = 1,2,\ldots$ . Assume that there is a measurable map $\varphi: \Omega_1 \to \Omega_2$ such that $T_2 \circ \varphi = \varphi \circ T_1$, $P_2 = \varphi(P_1)$ and there exists a continuous linear map $\Psi: V_1 \to V_2$ such that $Y_2 \circ \varphi = \Psi \circ Y_1$.

1.2 PROPOSITION. *If* $k: V_1 \to [0,\infty]$ *is an upper [lower] deviation function for* $Y_1$, *then* $\tilde{k}: V_2 \to [0,\infty]$ *defined by*

$$\tilde{k}(v_2) = \inf\{k(v_1): \ \Psi(v_1) = v_2, \ v_1 \in V_1\}$$

*with* $\tilde{k}(v_2) = \infty$ *if not otherwise defined, is an upper [lower] deviation function for* $Y_2$. *If* $k$ *is the deviation function for* $Y_1$ *then* $\tilde{k}^{\vee}$ *is the deviation function for* $Y_2$, *and if the compactness condition holds for* $(Z_n^{(1)})$ $\tilde{k}^{\vee} = \tilde{k}$ *and the compactness condition holds for* $(Z_n^{(2)})$.

The proof is straightforward and will not be given. The function $\tilde{k}$ will be referred to as the *induced deviation function*. Related results are given in [BZ] under the heading "pullback entropy" and in

[V] by the name "contraction principle."

We specialize now to the situation that will concern us for most of this work. Let $S = (\Omega, F, P, T)$ where $\Omega$ is a separable metrisable space with Borel sets $F$, $P$ is a probability measure on $(\Omega, F)$, and $T: \Omega \to \Omega$ is measurable. We call $S$ a *generalized dynamical system.* If $T$ is measure preserving, that is, if $P = TP$, we call $S$ a *dynamical system.*

Let $M$ or $M(\Omega)$ be the set of probability measures on $(\Omega, F)$ topologized by weak convergence. A measure $Q \in M$ is called *invariant* or stationary if $Q = TQ$ and $M_T$ denotes the class of stationary $Q$. Finally $M_{T,e}$ denotes the ergodic members of $M_T$.

Now let $\delta : \Omega \to M$ be defined by

$$\delta_\omega(A) = \begin{cases} 1, & \omega \in A, \\ 0, & \omega \notin A. \end{cases}$$

and let

$$L_n(\omega) = \frac{1}{n} \sum_{k=0}^{n-1} \delta_{T^k \omega}, \qquad n = 0,1,\ldots .$$

Our main concern will be the existence of a deviation function $K$ for the random variable $\delta$, that is for the sequence $(L_n)$. We shall say that a function $K : M \to [0,\infty]$ is an *(upper)* [*lower*] *deviation function* if it is an (upper) [lower] deviation function for $(L_n)$. Similarly $K_*$ and $K^*$ will have the same significance as in Proposition 1.1, but with respect to the sequence $(L_n)$. Observe that if $P \in M_T$ the ergodic theorem implies that $L_n(\omega)$ converges to a limit $\bar{P}(\omega) \in M_T$ $P$ - a.e., and if $P \in M_{T,e}$ then $\bar{P}(\omega) = P$, $P$ - a.e.

One motivation for studying this problem is as follows. Consider a bounded continuous (vector valued) random variable $Y$ on $(\Omega, F, P, T)$. A deviation function $K$ for our problem induces one for $Y$ as explained in Proposition 1.2, where $\varphi: \Omega \to \Omega$ is the identity and $\Psi$ is defined on $M$ by

$$\Psi(Q) = \int Y \, d Q .$$

So the existence of $K$ ensures the existence of deviation functions $k_Y$ for a large class of $Y$; indeed the class of $Y$ is much bigger than indicated, see Corollary 1.5. Another (related) motivation is given by an observation of Donsker and Varadhan presented below as Theorem 1.6.

For a generalized dynamical system the *deviation carrier* will be $\{Q \in M: \; K_*^{\wedge}(Q) < \infty\}$ .

1.3 PROPOSITION. *Let* $S = (\Omega, F, P, T)$ *be a generalized dynamical system and suppose* $T$ *is continuous. Then the deviation carrier is included in* $M_T$ .

PROOF. Let $Q \in M \setminus M_T$ . Then there exists a bounded continuous function $f$ on $\Omega$ such that

$$\alpha = \int f \, d Q \neq \int f \circ T \, dQ = \beta .$$

By hypothesis, $f \circ T$ is also continuous. For $\varepsilon > 0$ let

$$N_\varepsilon = \{Q' \in M: \; \left|\int f \, dQ' - \alpha\right| < \varepsilon, \quad \left|\int f \circ T \, dQ' - \beta\right| < \varepsilon\} .$$

Then,

$$[L_n \in N_\varepsilon] = [\left|\frac{1}{n} \sum_{k=0}^{n-1} f \circ T^k - \alpha\right| < \varepsilon \; , \; \left|\frac{1}{n} \sum_{k=1}^{n} f \circ T^k - \beta\right| < \varepsilon]$$

and for $\varepsilon < |\beta - \alpha|/2$ and $n$ sufficiently large this set will be empty. It follows that $K_*(Q) = \infty$ . Since $M_T$ is closed we can conclude

$K_*^\wedge(Q) = \infty$ , so that $Q$ is not in the deviation carrier. □

Note that for a generalized dynamical system the compactness condition is equivalent to the existence of a sequence of compact subsets $C_n$ of $M$ such that

(1.4) $$\lim_{n\to\infty} \overline{\lim_{m\to\infty}} \frac{1}{m} \log P[L_m \in M\setminus C_n] = -\infty .$$

Let $S = (\Omega,F,P,T)$ be a generalized dynamical system. A subset $\Omega_0$ of $\Omega$ will be called *negligible* if there exists $\Omega_1 \in F$ such that (i) $\Omega_0 \subseteq \Omega_1$ , (ii) $T(\Omega\setminus\Omega_1) \subseteq \Omega\setminus\Omega_1$ , (iii) $P(\Omega_1) = 0$ , (iv) $Q(\Omega_1) = 0$ for every $Q$ in the deviation carrier.

Consider two generalized dynamical systems. $S = (\Omega,F,P,T)$ and $S' = (\Omega', F',P',T')$ . Say that $S$ is homomorphic to $S'$ if there exists a negligible subset $\Omega_0$ of $\Omega$ and a mapping $\varphi$ of $\Omega\setminus\Omega_0$ into $\Omega'$ satisfying the following conditions

(a) $\varphi(P) = P'$

(b) $\varphi\circ T(\omega) = T'\circ\varphi(\omega)$, $\omega \in \Omega \setminus \Omega_0$

(c) $\varphi$ restricted to $\Omega\setminus\Omega_0$ is continuous in the relative topology.

(1.4) PROPOSITION. *Let* $S = (\Omega,F,P,T)$ *and* $S' = (\Omega',F',P',T')$ *be two generalized dynamical systems and* $\varphi$ *a homomorphism from* $S$ *to* $S'$. *For* $K: M(\Omega) \to [0,\infty]$ *define* $\tilde{K}: M(\Omega') \to [0,\infty)$ *by*

$$\tilde{K}(Q') = \inf\{K(Q):\varphi(Q) = Q'\}$$

*with* $\tilde{K}(Q') = \infty$ *if not otherwise defined.*

(i) *If* $K$ *is the deviation function of* $S$ , *then* $\tilde{K}^\wedge$ *is the deviation function for* S' .

(ii) *If* K *is an upper deviation function for* S *, then* $\tilde{K}$ *is an upper deviation function for* S' .

(iii) *if* K *is a lower deviation function for* S *that is finite only on the deviation carrier, then* $\tilde{K}$ *is a lower deviation function for* S' .

PROOF. IF K satisfies the hypothesis of (i) it satisfies the hypotheses of (ii) and (iii). It will suffice to prove (ii) and (iii).

Assume first that the negligible set entering in the definition of homomorphism is void. Then $\varphi$ is defined on $\Omega$ , and the induced mapping (still denoted by $\varphi$ ) of $M(\Omega)$ into $M(\Omega')$ is linear and continuous. As usual $\delta$ denotes the map $\Omega \to M(\Omega)$ carrying $\omega$ into the unit mass concentrated on $\{\omega\}$ ; let us denote the corresponding map $\Omega' \to M(\Omega')$ by $\delta'$ . Note $\varphi(\delta_\omega) = \delta_{\varphi(\omega)}$ , $\omega \in \Omega$ . So we may apply Proposition 1.2 (with $\varphi$ and $\Psi$ of that proposition corresponding respectively to $\varphi : \Omega \to \Omega'$ and $\varphi : M(\Omega) \to M(\Omega'))$ .

If the negligible set $\Omega_0$ is not void we may assume, referring to the definition of negligible set, that $\Omega_0 = \Omega_1$ , so that $\Omega \setminus \Omega_1$ is closed under T . Let $\hat{S}$ be the system obtained from S by restricting everything to $\Omega \setminus \Omega_1$ . Let K be a lower deviation function finite only on the deviation carrier of S , or an upper deviation function (in that case K is automatically finite only on the deviation carrier). All that needs showing is that when K is restricted to $\{Q: Q(\Omega \setminus \Omega_1) = 1\}$ it remains a lower, respectively upper, deviation function for $\hat{S}$ . But on the one hand those Q for which $Q(\Omega_1) > 0$ satisfy $K(Q) = \infty$, on the other hand for $N \subseteq M(\Omega)$

$$P[\omega \in \Omega : \frac{1}{n}(\delta_\omega + \delta_{T\omega} + \ldots \delta_{T^{n-1}\omega}) \in N]$$

is equal to

$$P[\omega \in \Omega \setminus \Omega_1 : \frac{1}{n}(\delta_\omega + \delta_{T\omega} + \dots \delta_{T^{n-1}\omega}) \in \hat{N}]$$

where $\hat{N} = N \cap \{Q \in M: Q(\Omega \setminus \Omega_1) = 1\}$ because $T(\Omega \setminus \Omega_1) = \Omega \setminus \Omega_1$ and $P(\Omega \setminus \Omega_1) = 1$. □

1.5 COROLLARY. *Let* S *be a generalized dynamical system,* Y *a real valued random variable on the probability space of* S *, and suppose* S *has a deviation function* K *. If* Y *is bounded and there exists a negligible set* $\Omega_0$ *such that* Y *restricted to* $\Omega \setminus \Omega_0$ *is continuous in the relative topology, then*

$$k_Y(y) = \inf\{K(Q): \ Q \in M, \ \int Y\, dQ = y\}$$

*is the deviation function for* Y .

PROOF. Suppose first that Y is bounded and everywhere continuous. Then

$$\Psi(Q) = \langle Y,Q\rangle = \int Y\, dQ$$

defines a bounded linear function on $M$ , and since

$$\frac{1}{n}\sum_{k=0}^{n-1} Y\circ T^k = \langle Y,L_n\rangle ,$$

the result in this case follows from Proposition 1.2, with $\varphi$ the identity, $\Psi$ as above.

Now assume Y is bounded and Y is continuous on $\Omega \setminus \Omega_0$ in the relative topology, where $\Omega_0$ is negligible. Again referring to the definition of negligible set, we may suppose $\Omega_0 = \Omega_1$. As in the proof of Proposition 1.4, we may restrict the system to $\Omega \setminus \Omega_1$ and K restricted to $M(\Omega \setminus \Omega_1)$ will provide a deviation function for the restricted system. On the restricted system Y is continuous, and the result follows from the first part of the proof. □

The functions $\tilde{K}$ and $k_Y$ in Proposition 1.4 and Corollary 1.5 will again be referred to as *induced* deviation functions.

We will use E[ ] for the expected value operator:

$$E[Y] = \int Y \, d \, P$$

When some other probability is involved it will appear as a superscript, e.g. $E^Q[Y] = \int Y \, d \, Q$. We shall also write $\langle Y,Q \rangle$ for $E^Q[Y]$.

Now we cite a basic fact about deviation functions, noted originally by Donsker and Varadhan and explaining the title of [DV]. The proof is straightforward; it can be found in a more general context in [V].

1.6 THEOREM. (Donsker and Varadhan). *Let* S *be a generalized dynamical system with deviation function* K *and* F *a real valued bounded continuous function on* M . *Then*

$$\lim_{n\to\infty} \frac{1}{n} \log E \exp\{n F(L_n)\} = \sup_Q (F(Q) - K(Q)) .$$

1.7 COROLLARY. *Let* S *be a generalized dynamical system with deviation function* K *and* Y *a bounded continuous real valued random variable on the probability space of* S . *Then*

$$C(Y) = \lim_{n\to\infty} \frac{1}{n} \log E[\exp\{ \sum_{k=0}^{n-1} Y\circ T^k\}]$$

*exists and satisfies*

$$C(Y) = \sup_{Q\in M} \{\langle Y,Q\rangle - K(Q)\} .$$

PROOF. In Theorem 1.6, let $F(Q) = \langle Y,Q\rangle$ . □

1.8 REMARKS. (i) In the language of convex functions the final equation in Corollary 1.7 says that $C$ is the *dual* of $K$ . This automatically makes $C$ convex and lower semicontinuous; see [BP]. A convex function is called *proper* if it is always greater that $-\infty$ and not identically $+\infty$ . If $K$ is a proper convex function and lower semicontinuous $K$ will be the dual of $C$ , i.e.

$$K(Q) = \sup_{Y} \{\langle Y,Q\rangle - C(Y)\} .$$

(ii) If in Corollary 1.7 one considers $Y$ which takes value in a Banach space, one may apply the Corollary and Remark (i) to $\xi(Y)$ , where $\xi$ is in the dual space. Then

$$c(\xi) \equiv C(\xi(Y)) = \lim_{n\to\infty} \frac{1}{n} \log E[\exp\{\xi( \sum_{k=0}^{n-1} Y\circ T^k)\}] .$$

Let

$$k(x) = k_Y(x) = \inf\{K(Q): Q \in M, \langle Y,Q\rangle = x\}$$

be the induced deviation function for $Y$ . Then

$$c(\xi) = \sup_Q\{\xi(<Y,Q>) - K(Q)\} = \sup_x\{\xi(x) - k(x)\}$$

so that $c$ is the dual of $k$. Whenever $k$ is a proper convex function one can obtain it by duality from $c$,

$$k(x) = \sup_\xi\{\xi(x) - c(\xi)\} .$$

In this form $k$ is sometimes called the Cramer transform, (see [Az]). Indeed consider the classical situation treated in [Cr] and [C]: there $Y$, $Y \circ T$, $Y \circ T^2, \ldots$ is a sequence of independent, identically distributed, real valued random variables and

$$c(\xi) = \log E[e^{\xi Y}] , -\infty < \xi < \infty .$$

Then the deviation function $k_Y$ for $Y$ is the dual of $c$,

$$k_Y(x) = \sup_\xi\{\xi x - c(\xi)\} .$$

Incidentally if $x \geq E[Y]$ one obtains (see[Az]) that

$$\frac{1}{n} \log P[Y + Y \circ T + \ldots + YT^{n-1} > x] < -k_Y(x) , n = 1,2\ldots .$$

Sometimes it is easy to calculate $k_Y$ explicitly. In case $P[Y = 1] = p = 1 - P[Y = 0]$, one obtains

$$k_Y(x) = x \log \frac{x}{p} + (1 - x)\log \frac{1 - x}{1 - p} , 0 \leq x \leq 1$$

where $0 \log 0 = 0$, and $k_Y(x) = \infty$ for $x \notin [0,1]$.

For explicit calculations of deviation functions by methods related to the present discussion see [E] and [CG].

(iii) In most work dealing with deviation functions they are indeed proper convex functions. This will be true in our work also. This seems to be due to the fact that one usually works with systems with good ergodic properties. For an instance of a deviation function that is not convex see Example 6.4.

(iv) In the Donsker-Varadhan theory [DV] and in some of our extensions $K$ is some kind of entropy function. In that case the duality relation in Corollary 1.7 between $C$ and $K$ corresponds to the duality between pressure and entropy in topological dynamics, see [DGS], Proposition 18.12.

(v) The mere existence of a deviation function does not rule out the possibility of certain degeneracies. Suppose for example that a dynamical system possesses a deviation function $K$ whose carrier is just a one-point set $\{P\}$. Then in fact we have super-exponential convergence. Notice that this is exactly what will happen if the dynamical system is uniquely ergodic. Another kind of degeneracy may be more troubling: the set of $Q$ on which $K$ vanishes may contain more than one point; of course for these $Q$ one does not have exponential convergence at all.

Such behavior is illustrated in Example 6.5. It appears to be due to the absence of good ergodic properties. For positive results see Proposition 2.3. Note that the hypothesis there is much weaker than Condition 5.8. The next remark is also relevant.

(vi) Suppose in (i) of Proposition 1.4 the deviation function $K$ has compact level sets. Then $\tilde{K}$ is lower semicontinuous. Also in that case $\tilde{K}(Q') = 0$ only if there exists a $Q$ such that $\varphi(Q) = Q'$ and $K(Q) = 0$.

## 2. Shifts and Entropy.

For every integer $n$, let $(M_n, \mathcal{B}_n) = (M, \mathcal{B})$, where $M$ is a complete separable metric space and $\mathcal{B}$ the corresponding Borel field. Let

$$(\Omega, F) = \prod_{n=-\infty}^{\infty} (M_n, \mathcal{B}_n)$$

be the product space, endowed with the product topology. It is again a complete separable metric space. The same applies to the one sided sequence spaces

$$(\Omega^+, F^+) = \prod_{n=1}^{\infty} (M_n, \mathcal{B}_n), \quad (\Omega^-, F^-) = \prod_{n=-\infty}^{0} (M_n, \mathcal{B}_n) .$$

Let $\theta$ be the shift on $\Omega$, defined by $(\theta\omega)_n = \omega_{n+1}$ for every integer $n$. The shift on $\Omega^+$ is defined similarly, with $n$ restricted to the positive integers. When the dependence on $M$ needs to be indicated we may write $\Omega(M)$, $\Omega^+(M)$, $\Omega^-(M)$. If $\omega = (\ldots\omega_{-1}\omega_0\omega_1\ldots) \in \Omega$, let $\omega^- = (\ldots\omega_{-1}\omega_0) \in \Omega^-$. For $P \in M(\Omega)$, $S = (\Omega, F, P, \theta)$ is a generalized dynamical system. It is called a shift or stochastic process. Note that $\theta: \Omega \to \Omega$ is continuous, so Proposition 1.3 applies. If $P \in M_\theta$ one obtains a dynamical system, also called in this case *stationary shift* or stationary stochastic process. The same remarks apply to systems $S^+ = (\Omega^+, F^+, P^+, \theta)$, where $P^+ \in M(\Omega^+)$ is the *one-sided shift*. If $S^+$ is a stationary one sided shift it can always be extended to a stationary shift; so frequently we will just work with the *bilateral shift*. Since $S$ and $S^+$ are completely determined by specifying $M$ and $P$ or $P^+$ we may write simply $S = [\Omega(M), P]$, $S^+ = [\Omega^+(M), P^+]$.

For $-\infty \leq m < n \leq \infty$, let $F_{m,n} = \Pi_{i=m}^{n} F_i$. If $P$ is any probability measure on $(\Omega, F)$, then $P_{m,n}$ will denote the restriction to $F_{m,n}$, and $P_m$ the restrictions to $F_m$ . $P \in M_\theta$ implies that P - a.e.

$$(2.1) \qquad P[A|F_{m,n}](\omega) = P[\theta^{-1}A|F_{m+1,n+1}](\theta\omega), \quad m < n \ , \ A \in F \ .$$

The condition (2.1) is weaker that stationarity; when it holds we shall say that P is *homogeneous*; (consider, for example, Markov processes with stationary transition probabilities). We now define $M'_\theta$ to denote the class of homogeneous probabilities.

For each $A \in F$, the random variable $P[A|F_{-\infty,0}]$ is defined only up to a P-null set. Under our assumptions there exist *regular conditional* probabilities, i.e. the choice of random variables can be made so that for each $\omega$, $P[\cdot|F_{-\infty,0}](\omega)$ is a probability measure on $(\Omega,F)$. We shall make such a choice and denote it by $P^*_\omega$ so that $P^*_\cdot[A]$ is a version of $P[A|F_{-\infty,0}](\cdot)$ . If $P \in M'_\theta$ we will require

$$(2.2) \qquad P_\omega[A] = P[\theta^{-1}A|F_{-\infty,1}] \ , \ A \in F, \ \omega \in \Omega \ .$$

However, the collection of probability measures $P^* = (P^*_\omega, \omega \in \Omega)$ is not uniquely determined by P, and we keep the superscript to recall this non-uniqueness. Since $P^*_\omega$ depends on $\omega$ only through $\omega^-$ we also write $P^*_{\omega-} = P_\omega$. In case $P^*_{\omega-}$ depends on $\omega^-$ only through $\omega_0$, i.e. $P^*_{\omega-} = P^*_{\omega_0}$ , $P^*$ is said to be *Markovian.*

Consider now a shift $(\Omega(M),P)$. Define maps $\Phi_1: \Omega(M) \to \Omega(\Omega^-(M))$, $\Phi_2: \Omega(\Omega^-(M) \to \Omega^+(\Omega^-(M))$ and $\Phi_3: \Omega^+(\Omega^-(M) \to \Omega(M))$ by the following procedure. For $\omega = (\omega_n) \in \Omega(M)$ let $\Phi_1(\omega) = (\omega^*_n)$ , where $\omega^*_n = (\ldots\omega_{n-1}\omega_n)$ , $n = 0,\pm1,\ldots$ . Let $\Phi_2$ send the bilateral sequence $(\omega^*_n, n = 0,\pm1,\ldots)$ into the one sided sequence $(\omega^*_n, n = 1,2\ldots)$. Finally this one sided sequence is mapped by $\Phi_3$ into the bilateral sequence $(\omega_n)$ , so that $\Phi_3 \circ \Phi_2 \circ \Phi_1$ is the identity. Then $\Phi_1, \Phi_2, \Phi_3$ is a sequence of homomorphisms such that

$$(\Omega(M),P) \xrightarrow{\Phi_1} \Omega(\Omega^-(M),\Phi_1(P)) \xrightarrow{\Phi_2} \Omega^+(\Omega^-(M),\Phi_2\circ I_1(P))$$
$$\xrightarrow{\Phi_3} (\Omega(M),P)$$

This shows how we can transfer results about deviation functions from two sided shifts to one sided shifts. Let us focus now on the shift $(\Omega(\Omega^-(M),\Phi_1(P))$ . The mapping $\Phi_1$ takes $\omega$ into $\omega^*$ and $\omega^*$ has the property that knowledge of $\omega_0^*$ implies knowledge of $\omega_{-1}^*,\omega_{-2}^*,\ldots$ . This suggests that this shift should be Markovian. Indeed if $P^*$ is a family of regular conditional probabilities associated with $P$ , we may set

$$\Phi_1(P)[A|F_{-\infty,0}(\Omega(\Omega^-(M)))] = P^*_{\omega_0^*}[A] , \quad A \in F_{0,\infty}(\Omega(\Omega^-(M)))$$

so that we may regard this process as being Markovian. We note however that the choice of $P^*$ is still arbitrary. In effect, what this procedure does is to allow us to associate with a given shift a Markovian shift, which depends however on the choice of $P^*$ .

Given a shift $(\Omega(M),P)$, with $P \in M_\theta'$ we wish to define a mapping $H_P\colon M \to [0,\infty]$ , "the entropy function associated with P." Unfortunately the mapping that we define will depend on $P^*$, and we will write $H_{P^*}$ . Our procedure will be to show that if $P^*$ can be chosen to satisfy certain conditions, then $H_{P^*}$ agrees with some function depending only on $P$ , and hence for all such choices of $P^*$ the functions $H_{P^*}$ agree. Our approach follows that of [P], but there the lack of uniqueness was not discussed. If one starts with a Markov process and begins by specifying the $P^*$ there will be no ambiguity; this is the context in which entropy is developed by Donsker and Varadhan [DV].

Consider now $P \in M'_\theta$ with associated regular conditional probabilities $P^* = (P_{\omega^-}, \omega^- \in \Omega^-)$ and $Q \in M$. Define $[QP^*] \in M$ by

$$[QP^*](A) = \int P^*_{\omega^-}(A) Q(d\omega^-)$$

(Roughly, this measure agrees with $Q$ up to time $0$, after that it proceeds according to $P^*$).

We turn to the definition of entropy. If $(B,\mathcal{B})$ is a measurable space and $\mu$ and $\nu$ two probability measures on it, the *entropy of* $\nu$ *with respect to* $\mu$ is defined by

$$(2.3) \qquad h_\mu(\nu) = \begin{cases} \int (\log \frac{d\nu}{d\mu}(x))\nu(dx), & \text{if } \nu \ll \mu \\ \infty, & \text{otherwise} \end{cases}$$

where $\ll$ is written to denote absolute continuity. Donsker and Varadhan [DV, Part 1], have given the following important variational characterisation of entropy:

$$(2.4) \qquad h_\mu(\nu) = \sup_{\Phi \in \mathcal{B}_b} \left(\int_B \Phi \, d\nu - \log \int_B e^\Phi d\mu\right)$$

where $\mathcal{B}_b$ denotes the class of bounded B-measurable functions. Furthermore they show that if $B$ is a separable complete metric space with Borel sets $\mathcal{B}$, the supremum may be taken over the class of bounded continuous functions.

For $P \in M'_\theta$ and $P^* = (P_\omega)$ an associated family of regular conditional probabilities, define the *entropy*

$$H_{P^*}(Q) = \begin{cases} h_{[QP^*]_{-\infty,1}}(Q_{-\infty,1}), & Q \in M_\theta \\ \infty, & Q \in M \setminus M_\theta \end{cases}$$

Recall that the subscript $-\infty,1$ denotes restriction to $F_{-\infty,1}$. For

$Q \in M_\theta$ one can express $H_{P*}(Q)$ more explicitly. Namely, let $Q^* = (Q_\omega)$ be a family of regular conditional probabilities associated with $Q$ and let $Q_\omega|_1$ be the restriction of $Q_\omega$ to $F_1$; similarly $P^*_\omega|_1$ denotes the restriction of $P^*_\omega$ to $F_1$. Then

$$(2.5) \qquad H_{P*}(Q) = \int_{\Omega^-}[\int_M \log \frac{dQ^*_{\omega^-}|_1}{dP^*_{\omega^-}|_1}(y) Q^*_{\omega^-}(dy)] Q(d\omega^-) .$$

Note that the choice of $P^*$ is unique up to a P-null set in $F_{-\infty,0}$, while the choice of $Q^*$ is unique up to a $\hat{Q}$-null set in $F_{-\infty,0}$. Since the definition of $H_{P*}(Q)$ involves an integration over $\Omega^-$ with respect to $Q$, the non-uniqueness in the choice of $Q^*$ is immaterial, and that is why the left side of (2.5) depends on $Q$, not $Q^*$. For this reason we will in the future, when writing formulas like (2.5) write $Q_\omega$ in place of $Q^*_\omega$.

As explained above, once a choice of $P^*$ has been made one is essentially in the Markovian situation, and the results of [DV] apply. We summarize some of these in the following theorem, referring for the proofs to [DV] or [V].

2.1 THEOREM. (Donsker and Varadhan). *With* $P^*$ *fixed let* $H$ *denote the function* $H_{P*}(\cdot)$ *restricted to* $M_\theta$.

(i) $H$ *is affine.*

(ii) *If the mapping from* $\Omega_-$ *to* $M$ *taking* $\omega^-$ *into* $P^*_{\omega-}$ *is continuous, then* $H$ *is lower semicontinuous.*

(iii) $h_{[QP^*]_{-\infty,n}}(Q_{-\infty,n}) = n\, H_{P*}(Q)$, $n = 1,2\ldots$ .

(iv) *If* $H(Q) < \infty$, *there exists a sequence of* $Q_n$ *such that each* $Q_n$ *is a finite convex combination of members of* $M_{\theta,e}$ *and both* $Q_n \to Q$ *in* $M$ *and* $H(Q_n) \to H(Q)$ *as* $n \to \infty$.

2.2 REMARKS. Referring to the previous theorem, note that even if $H$ is not lower semicontinuous its lower regularisation $\hat{H}$ will be. Furthermore, since $H$ is affine $\hat{H}$ will be a convex function.

2.3 PROPOSITION. *Let* $S = (\Omega, F, P, \theta)$ *be a shift,* $P \in M_{\theta,e}$ *with associated regular conditional probabilities* $P^*$ . *Suppose*

$$\rho \equiv \inf_{m\geq 0} \sup_{A\in F_{m,\infty}} \sup_{\bar{\omega}\in\bar{\Omega}} \inf_{\bar{\eta}\in\bar{\Omega}} [P_{\bar{\omega}}(A) - P_{\bar{\eta}}(A)] < 1$$

*If* $Q \in M_\theta$ *and* $H_{P^*}(Q) = 0$ , *then* $Q = P$ .

PROOF. Note that for $Q \in M_\theta$ , $H_{P^*}(Q) = 0$ if and only if

(a) $$Q_{\bar{\omega}} = P_{\bar{\omega}} \quad \text{for } Q\text{ - a.e. } \bar{\omega}$$

as can be seen from Theorem 2.1. If (a) holds, the exceptional set can be taken to be empty by redefining $Q_{\bar{\omega}}$ on this set. If $H_{P^*}(Q) = 0$ and $Q \neq P$ holds for some $Q \in M_\theta$ , these conditions will hold for some $Q \in M_{\theta,e}$ . But then $Q$ and $P$ are orthogonal. For $\varepsilon > 0$ there exists $N > 0$ and $A \in F_{-N,N}$ with $P(A) < \varepsilon$ , $Q(A) > 1 - \varepsilon$ . By stationarity

(b) $$P[\theta^{N+m}A] < \varepsilon$$

(c) $$Q[\theta^{N+m}A] > 1 - \varepsilon$$

By (a) and (c) there exists for every $m \geqq 0$ an $\bar{\omega}$ such that

(d) $$Q_{\bar{\omega}}[\theta^{N+m}A] = P_{\bar{\omega}}[\theta^{N+m}A] > 1 - \varepsilon$$

and by (b), for every $m \geqq 0$ there exists $\bar{\eta}$ such that

(e) $$P_{\bar{\eta}}[\theta^{N+m}A] < \varepsilon .$$

Choosing $\varepsilon$ so that $1 - 2\varepsilon > \rho$ , (d) and (e) are inconsistent. □

## 3. Upper Bounds for Shifts

Consider now a shift $S = (\Omega(M), F, P, \theta)$ as introduced in the previous section. We will assume $P \in M_\theta$ or $P \in M'_\theta$ .

Recall that $\delta_\omega$ is the probability measure concentrated on $\omega$ and

$$L_n(\omega) = \frac{1}{n} \sum_{k=0}^{n-1} \delta_{\theta^k \omega}$$

Define $\pi_n: \Omega \to \Omega$ by $(\pi_n\omega)_i = \omega_i$ , $0 \leq i < n$, and $(\pi_n\omega)_{i+n} = (\pi_n\omega)_i$ for all $i$ . Now let $L'_n = L_n \circ \pi_n$ . Donsker and Varadhan in [DV, Part 4] are concerned with deviation functions for the sequence $(L'_n)$ rather than $(L_n)$ . Observe that $L'_n$ takes values in $M_\theta$ . Write $L'_n(\omega,A)$ for $L'_n(\omega)(A)$ , and $L_n(\omega,A)$ for $L_n(\omega)(A)$ . For any bounded random variable $Y$ ,

$$\int Y \, dL_n = \frac{1}{n} \sum_{j=0}^{n-1} Y \circ \theta^j , \quad \int Y \, dL'_n = \frac{1}{n} \sum_{j=0}^{n-1} Y \circ \theta^j \circ \pi_n$$

and if $Y$ is also $F_{0,m}$-measurable

(3.1) $$\sup_\omega \left| \int Y(\eta) \, L_n(\omega,d\eta) - \int Y(\eta) \, L'_n(\omega,d\eta) \right| \leq \frac{2m}{n} \sup_\eta Y(\eta) .$$

Now $M$ is metrisable and one can show that the distance between $L_n(\omega)$ and $L'_n(\omega)$ converges to zero as $N \to \infty$ , uniformly in $\omega$ . To verify this one need only check that for any bounded random variable $Y$ the left side of (3.1) converges to zero. It is sufficient to consider $Y$

which are $F_{-m,m}$-measurable, m=1,2,... . Then $Y\circ\theta^k$ is $F_{-m+k,m+k}$-measurable, and so the desired result can be seen to follow from (3.1).

Associated with our shift S , are the functions $K^*$ and $K_*$ defined like $k^*$ and $k_*$ in Proposition 1.1 but with the sequence $(L_n)$ taking the place of $(Z_n)$. From the facts established about $L_n$ and $L'_n$ we see that if we let $(L'_n)$ play the role of $(Z_n)$ we would get the same $K^*$ and $K_*$ . Observe that the sequence $(L'_n)$ satisfies the compactness condition (1.3) if and only if there is a sequence $(C_n)$ of compact subsets of $M$ such that

$$\text{(3.2)} \qquad \lim_{n\to\infty} \overline{\lim_{m\to\infty}} \frac{1}{m} \log P[L'_m \in M \setminus C_n] = -\infty .$$

We regard $L'_n$ as an $M$-valued random variable, even though $L'_n$ takes its values in the subset $M_\theta$ .

3.1 PROPOSITION. (i) *If* K *is an upper deviation function for* $(L'_n)$ *it is an upper deviation function for* $(L_n)$ .

(ii) *If* K *is a lower semicontinuous lower deviation function for* $(L'_n)$ *and* K *has compact level sets, then* K *is a lower deviation function for* $(L_n)$ .

PROOF. Suppose K is an upper deviation function for $(L'_n)$ . Let $Q \in M$ and $N$ a neighborhood of Q . Since the distance between $L_n(\omega)$ and $L'_n(\omega)$ converges to zero uniformly in $\omega$ as $n \to \infty$ , there exists a neighborhood $N_0$ of Q such that for all sufficiently big n

$$P[L_n \in N] \geq P[L'_n \in N_0]$$

and hence

$$\underline{\lim}_{n\to\infty} \frac{1}{n} \log P[L_n \in N] \geq \underline{\lim}_{n\to\infty} \frac{1}{n} \log P[L_n' \in N_0] \geq -K(Q)$$

and this implies that $K$ is an upper deviation function for $(L_n)$.

Now assume that $K$ satisfies the hypotheses of (ii). Let $N$ be a measurable subset of $M$ and $N^{(\varepsilon)}$ an $\varepsilon$-neighborhood of $N$. Then

$$-a_0 \equiv \overline{\lim}_{n\to\infty} \frac{1}{n} \log P[L_n \in N] \leq \overline{\lim}_{n\to\infty} \frac{1}{n} \log P[L_n' \in N^{(\varepsilon)}]$$

$$\leq -\inf\{K(Q): Q \in \overline{N^{(\varepsilon)}}\} \equiv -b_\varepsilon \ .$$

We assume $-\infty < -a_0$ (otherwise there is nothing to prove), so that $\infty > a_0 \geq b$. By assumption $K$ is greater than $a_0$ outside some compact set $C$ so that

$$b_\varepsilon = \inf\{K(Q): Q \in N^{(\varepsilon)} \cap C\},$$

and by the lower semicontinuity of $K$, $b_\varepsilon = K(Q^\varepsilon)$ for some $Q^\varepsilon \in N^{(\varepsilon)} \cap C$. Taking a sequence of $\varepsilon_n \downarrow 0$ we obtain corresponding $Q^{\varepsilon_n}$, and, taking a subsequence if necessary, we may assume $Q^{\varepsilon_n} \to Q^0 \in \bar{N}$ as $n \to \infty$. Again by lower semicontinuity one obtains

$$K(Q^0) = \lim_{n\to\infty} b_{\varepsilon_n} = \inf\{K(Q): Q \in \bar{N}\}$$

and $-a_0 \leq -K(Q^0)$ as desired. □

Consider a shift $S = (\Omega, F, P, \theta)$, $P \in M_\theta'$, and $P^*$ a choice of regular conditional probabilities. The following condition (which was introduced in (ii) of Theorem 2.1) will be important:

(3.3) The map from $\Omega^-$ to $M$ taking $\omega^-$ into $P^*_{\omega^-}$ is continuous.

We note that if we go over to the Markovian shift associated with $P^*$, (3.3) is a condition assumed in [DV, part 4]. In the present discrete parameter context this condition is equivalent to the Markov process having the Feller property.

3.2 LEMMA (Donsker and Varadhan). *Let* $S = (\Omega, F, P, \theta)$ *be a shift,* $P \in M'_\theta$ *and* $P^*$ *a choice of regular conditional probabilities satisfying* (3.3). *For any compact subset* $N$ *of* $M$

$$\overline{\lim_{n\to\infty}} \frac{1}{n} \log P^*_{\omega^-}[L'_n \in N] \leq \inf\{H_{P^*}(Q): Q \in N\}$$

*and the convergence is uniform in* $\omega^-$.

PROOF. This is essentially Lemma 4.3 of [DV, part 4], or Lemma 11.3 in [V]. As explained in Section 2, once $P^*$ is chosen we can consider our shift to be Markovian by using the map $\Phi_1(\omega) = \omega^*$, where $\omega^*_n = (\ldots\omega_{n-1}, \omega_n)$.

3.3 THEOREM. *Let* $S = (\Omega, F, P, \theta)$ *be a shift,* $P \in M'_\theta$ *and* $P^*$ *a choice of regular conditional probabilities satisfying* (3.3). *Assume* (3.2) *holds for* $L'$. *Then* $H_{P^*}$ *is a lower deviation function for* $(L'_n)$ *and for* $(L_n)$.

PROOF. By Lemma 3.2 and (v) of Proposition 1.1, $H_{P^*}$ is a lower deviation function for $(L'_n)$. The assertion for $(L_n)$ follows from (ii) of Theorem (2.1) and (ii) of Proposition 3.1.

If, in Theorem 3.3, $\Omega = \Omega(M)$ with $M$ compact, then $M$ is a compact set and the compactness condition for $(L_n')$ holds automatically. However if $M$ is not compact one needs a verifiable condition on the shift that implies that $(L_n')$ satisfies the compactness condition. Such a condition - though a very stringent one - is given in the next theorem.

3.4 THEOREM. *Let* $S = (\Omega(M), F, P, \theta)$ *be a shift* $P \in M_\theta'$. *Suppose there exists a sequence* $(M_n)$ *of compact subsets of* $M$ *such that*

$$P[\omega_1 \notin M_n | F_{-\infty,0}] \leq 2^{-n} \quad P\text{ - a.e.}$$

*Then the compactness condition* (3.2) *for* $(L_n')$ *is satisfied.*

PROOF. If $(\Omega_n)$, $n = 0,1,\ldots$ is a sequence of compact subsets of $\Omega$,

$$C(\Omega.) = \{Q \in M\colon Q(\Omega \setminus \Omega_n) < 2^{-n}, \ n = 0,1,\ldots\}$$

defines a compact subset of $M$. If $(M_n)$ is a sequence of compact subsets of $M$,

$$C'(M.) = \{Q \in M_\theta\colon Q_i(M \setminus M_n) < 8^{-n}, \ n = 0,1,\ldots\}$$

will be shown to define a compact subset of $M_\theta$; (recall that $Q_i$ is $Q$ restricted to $F_i$, and since $Q \in M_\theta$ the definition is independent of i). Indeed setting

$$\Omega_n(M.) = \{\omega\colon \ \omega_i \in M_{n+|i|}, \ i = 0,\pm 1,\ldots\}, \ n = 0,1,\ldots$$

one finds that $C'(M.) \subseteq C(\Omega.(M.))$, and since $C'(M)$ is closed, it is compact.

Now assume that $(M_n)$ satisfies the condition of the Theorem. We have to construct $(C_n)$ so that (3.2) holds. For each $n$, $(s(N,k)\ ,\ k = 0,1,\dots)$ will be a suitable sequence of positive integers, and $C_n = C'(M_{s(n,\cdot)})$. Then

$$[L'_m \notin C'(M_{s(n,\cdot)})] = [(L'_m)_k(M \setminus M_{s(n,k)}) > 8^{-|k|} \quad \text{for some} \quad k]$$

and so

$$P[L'_m \notin C'(M_{s(n,\cdot)})] \le 2 \sum_{k=0}^{\infty} P[(L'_m)_1(M \setminus M_{s(n,k)}) > 8^{-k}] .$$

But note that for any $N \in \mathcal{B}$,

$$(L'_m)_1(N) = \frac{1}{m} \sum_{k=0}^{m-1} \chi_N(\omega_k)$$

and the hypothesis of the theorem ensures that for each $M_i$, each positive integer $m$ and $t > 0$,

$$P[\sum_{k=0}^{m-1} \chi_{M \setminus M_i}(\omega_k) > t] \le P[Y_1^{(i)} + \dots + Y_m^{(i)} > t]$$

where $Y_1^{(i)}, Y_2^{(i)}, \dots$ form a sequence of independent identically distributed random variables with $P[Y_1^{(i)} = 1] = 2^{-i}$ and $P[Y_1^{(i)} = 0] = 1 - 2^{-i}$. We have then

$$P[L'_m \notin C'(M_{s(n,\cdot)})] \le 2 \sum_{k=0}^{\infty} P[\frac{1}{m}(Y_1^{s(n,k)} + \dots + Y_m^{s(n,k)}) > 2^{-k}]$$

One needs to observe that if $b$ is any positive number, the kth term on the right can be made less than $\exp\{-mb\}$ by choosing $s(n,k)$ sufficiently big. Though this follows from elementary estimates on tails of binomial distributions we have all the information necessary in (iii) of Remarks 1.8. It follows that for any positive numbers $b_n$ , one can insure that the right side of the last inequality is bounded above by $\exp\{-mb\}$ . Taking a sequence $(b_n)$ with $b_n \to \infty$ , we obtain the desired relation (3.2). □

## 4. Shannon-McMillan Theorems

Consider a shift as in Section 2. For $Q,P \in \mathcal{M}$ define

$$Q \ll_{loc} P \quad \text{if} \quad Q_{m,n} \ll P_{m,n} , \qquad -\infty < m \leq n < \infty.$$

Generalized Shannon-McMillan theorems start with $P \in \mathcal{M}_\theta'$ , associated regular conditional probabilities $P^*$, and $Q \in \mathcal{M}_\theta$ such that $Q \ll_{loc} P$, and they assert

$$\lim_{n\to\infty} \frac{1}{n} \log \frac{dQ_{on}}{dP_{on}} = Z , \qquad \int Z\, dQ = H_{P^*}(Q) \tag{4.1}$$

where the convergence is in Q-measure, or in $L'(Q)$ . Evidently if $P^*$ and $P^{**}$ are two choices of regular conditional probabilities for which (4.1) holds then $H_{P^*}(Q) = H_{P^{**}}(Q)$ .

The classical Shannon-McMillan theorem treats the case where the shift is on a finite space, i.e. M has only finitely many elements, P is product measure, and $P_1$ assigns equal weight to all elements of M. Following earlier work of Perez, Moy in [M] obtained a Shannon-McMillan theorem for shifts where P is Markovian, M being quite general. We give an extension in Theorem 4.2. Professor Perez has informed us of some related work $[P_1]$ (in which other references may be found). Our proof of Theorem 4.2 follows [M] rather closely. We shall omit the proof.

We hope to deal with related questions elsewhere. The Condition 4.1 which is assumed in the Theorem 4.2 may seem undesirable, but counter-examples presented by Kieffer [K] show the need for care, and I doubt that the condition can be much improved.

Let $S = [\Omega(M), P]$ be a shift with $P \in M'_\theta$. For $-\infty < m < n < \infty$ let $P[\ |F_{m,n}]$ be a choice of regular conditional expectations such that for every $A \in F$, $P[A|F_{m,n}] = P[\theta^k A|F_{m+k,n+k}]$ holds everywhere, $k = 1,2,\ldots$. Let $P_k[\ |F_{m,n}]$ denote the restriction of $P[\ |F_{m,n}]$ to $F_k$.

The following condition is satisfied in the Markovian case.

4.1 CONDITION. $P \in M'_\theta$ *and for every non-negative integer* $n$ *and every* $\omega$, $P_1[\ |F_{-n-1,0}](\omega) \ll P_1[\ |F_{-n,0}](\omega)$ *with strictly positive Radon-Nikodym derivative* $(1 + \alpha_n(y,\omega))$, $y \in M_0$ *such that* $\alpha_n^* = \sup_{\omega,y} \alpha_n(y,\omega)$ *satisfies* $\sum_{n=1}^{\infty} \alpha_n^* < \infty$.

Observe that in Condition 4.1 the convergence of $\sum \alpha_n^*$ is equivalent to the product $\prod_{n=0}^{\infty} (1 + \alpha_n(y,\omega))$ converging uniformly in $y$ and $\omega$, and this implies that $P_1[\ |F_{-n-m,0}]$ and $P_1[\ |F_{-n,0}]$ are mutually absolutely continuous with

$$P_1[dy|F_{-n-m,0}](\omega) = \prod_{k=n}^{n+m-1} (1 + \alpha_k(y,\omega))P_1[dy|F_{-n,0}](\omega) , \tag{4.2}$$

for $0 < n < \infty$, $0 < m < \infty$. Furthermore, we may let $m = \infty$ in (4.2). By assumption the right side converges uniformly in $\omega$, and the left-side converges to $P_1(dy|F_{-\infty,0})$ $P$ - a.e. by the martingale convergence theorem. Thus (4.2) serves to pick out a good version of $P_1[\ |F_{-\infty,0}](\omega) = P^*_{\omega^-}[\omega_1 \in \cdot]$. It follows from Theorem 4.2 that all $P^*$ obtained in this fashion give rise to the same entropy function $H_{P^*}(Q)$.

For $Q \in M$, define

$$[QP]_{m,n}(A) = \int P[A|F_{m,n-1}](\omega)\, Q_{m,n}(d\omega)\ , \quad A \in F_{m,n}\ , \ m < n\ .$$

4.2 THEOREM. *Let* $P \in M'_\theta$ *satisfy Condition* 4.1 *and let* $P^*$ *be obtained as above. For any* $Q \in M_\theta$ *satisfying* $Q \ll_{loc} P$ *and* $H_{P^*}(Q) < \infty$ *the assertion* (4.1) *holds, with the limit in the sense of Q-measure.*

*Furthermore, if* $r_n$ *is the Radon-Nikodym derivative of* $[Q(\ |F_{n,0})]_1$ *with respect to* $[P(\ |F_{n,0})]_1$ *then* $H^n = E^Q[\log r_n]$ *converges to* $H_{P^*}(Q)$ *as* $n$ *approaches infinity.*

*If in addition the Radon-Nikodym derivative of* $Q_0$ *with respect to* $P_0$ *is in* $L^1(Q)$ *the convergence in* (4.1) *holds in the* $L^1(Q)$ *sense.*

4.3 REMARK. The theorem does not allow us to infer that $H_{P^*}(Q) < \infty$ implies $Q \ll_{loc} P$. This can fail even when $P$ is Markovian, see Example 6.3(a).

## 5. Lower Bounds for Shifts

For $S = (\Omega, F, P, \theta)$ a shift, $P \in M'_\theta$ we discuss two approaches for obtaining lower bounds, that is for obtaining upper deviation functions. It follows from (i) of Proposition 3.1 that $K$ will be an upper deviation function for $S$ (i.e. for the sequence $(L_n)$) if for every $Q \in M_\theta$ and every neighborhood $N$ of $Q$

$$\underline{\lim}\ \frac{1}{n} \log P[L'_n \in N] \geq -K(Q)\ . \tag{5.1}$$

The first approach is based on the Shannon-McMillan theorem. Under certain conditions we obtain (5.1) for $Q \in M_{\theta,e}$ ; under more stringent conditions we obtain the result for all $Q \in M_\theta$ .

5.1 PROPOSITION. *Assume* $P \in M'_\theta$ *satisfies condition* 4.1 *and let* $P^*$ *be as in Theorem* 4.2. *Let* $Q \in M_{\theta,e}$ *with* $Q \ll_{loc} P$ , *and let* $N$ *be a neighborhood of* $Q$ . *Then* (5.1) *holds with* $K(Q) = H_{P^*}(Q)$ .

PROOF. Let

$$f_n = \frac{dQ_{0,n-1}}{dP_{0,n-1}} .$$

Since $L'_n$ is $F_{0,n-1}$-measurable, one obtains for $\varepsilon > 0$

$$P[L'_n \in N] \geq \int_{L'_n \in N} \exp(-\log f_n)\, dQ$$

$$\geq \exp[-n(H_{P^*}(Q) + \varepsilon)]\; Q[L'_n \in N, \frac{1}{n} \log f_n \leq H_{P^*}(Q) + \varepsilon]$$

By the ergodic theorem and theorem 4.2 the second factor in the last member converges to 1 as $n \to \infty$ . □

Proposition 5.1 gives information only for $Q \in M_{\theta,e}$ . To deal with $Q \in M_\theta$ we need to strengthen condition 4.1 and add an independence condition.

5.2 CONDITION. $P \in M'_\theta$ *and the Condition* 4.1 *holds and also* $P_1[\;|F_0](\omega) \ll P_1$ *with Radon-Nikodym derivative* $1 + \alpha(y,\omega)$ , *and there exists a positive constant* $m_1$ *such that*

$$\frac{1}{m_1} \leq 1 + \alpha(y,\omega) \leq m_1 \ , \qquad (y,\omega) \in M \times \Omega$$

$$\frac{1}{m_1} \leq 1 + \alpha_k(y,\omega) \leq m_1 \ , \qquad k = 0,1,\ldots, \ (y,\omega) \in M \times \Omega$$

5.3 REMARK. Condition 5.2 ensures that $P_1[\ |F_{\infty,0}] \ll P_1$ with Radon-Nikodym derivative $(1 + \alpha)\Pi(1 + \alpha_k)$ bounded uniformly above and below by positive constants.

The following lemma is known, but we prove it because we shall rely on the particular construction given in the proof.

5.4 LEMMA. *Let* $Q \in M_\theta$ . *Then there exist* $Q^N \in M_{\theta,e}$ , $N = 1,2,\ldots$ *such that* $Q^N \to Q$ *as* $N \to \infty$ .

PROOF. We first specify an auxiliary measure $Q^{N,0} \in M$ by specifying

(i) $Q^{N,0}_{1,N} \equiv (Q^{N,0})_{1,N} = Q_{1,N}$

(ii) $\theta^N Q^{N,0} = Q^{N,0}$ , (recall $\theta^N Q^{N,0} \equiv Q^{N,0} \circ \theta^{-N}$)

(iii) Under $Q^{N,0}$ the sequence of $\sigma$-fields $\ldots F_{N+1,0}, F_{1,N}, F_{N+1,2N}, \ldots$ is independent.

For any integer $k$ let $Q^{N,k} = \theta^k Q^{N,0}$ . As a function of $k$ , $Q^{N,k}$ has period $N$ , but the least period could be less than $N$ . Now let

$$Q^N = \frac{1}{N}\sum_{k=0}^{N-1}\theta^k Q^{N,0}$$

Evidently $Q^N \in M_\theta$ , and it is easy to see that $Q^N \to Q$ . Condition

(iii) ensures that each $Q^{N,k}$ satisfies the Kolmogorov zero-one law. (i.e. $A \in \bigcap_n F_{n,\infty}$ implies $Q^{N,k}(A)$ equals zero or one.) Suppose now $B \in F$, $\theta^{-1}B = B$. Then $B \in \bigcap_n F_{n,\infty}$ and so $Q^{N,k}(B) = b_k \in \{0,1\}$. Since $b_{k+1} = Q^{N,k+1}(B) = Q^{N,k}(\theta^{-1}B) = Q^{N,k}(B) = b_k$, $b_k = b_0$ for all $k$, and $Q^N(B) = b \in \{0,1\}$, showing $Q^N \in M_{\theta,e}$. □

5.5 THEOREM. *Let* $P \in M_\theta'$ *satisfy Condition* 5.2 *and let* $P^*$ *be as in Theorem* 4.2. *Let* $Q \in M_\theta$ *satisfy* $H_{P^*}(Q) < \infty$. *Then* $Q \ll_{loc} P$ *and there exists a sequence* $Q^N$, $N = 1,2,\ldots$ *with* $Q^N \in M_{\theta,e}$ *such that* $Q^N \to Q$ *and* $H_{P^*}(Q^N) \to H_{P^*}(Q)$ *as* $N \to \infty$.

PROOF. Since $H_{P^*}(Q) < \infty$, it follows from (iii) of Theorem 2.1 that if $A \in F_{1,n}$ then $Q_\omega(A) > 0$ implies $P^*_\omega(A) > 0$, for $Q$ - a.e. $\omega$. Assume $Q(A) > 0$, then certainly $P^*_\omega(A) > 0$ for some $\omega$. It follows from Condition 5.2 that $P^*_\omega(A) > 0$ if and only if the product measure $(P_1 \times P_1 \times \ldots P_1)$ on $F_{1,n}$ assigns positive measure to $A$. Hence $P^*_\omega(A) > 0$ for some $\omega$ implies the same relation for all $\omega$, hence $P(A) > 0$. Since $Q \in M_\theta$ and $P \in M_\theta'$ one can prove in the same manner that $Q_{-m,-m+n} \ll P_{-m,-m+n}$ for $m = 1,2,\ldots$; so $Q \ll_{loc} P$.

Now the sequence $Q^N$ is to be chosen as in the proof of Lemma 5.4. It must be shown that $H_{P^*}(Q^N) \to H_{P^*}(Q)$.

For $\omega \in \Omega(M)$ define $\omega^N = \Psi_N(\omega) \in \Omega(M^N)$ by

$$\omega_i^N = (\omega_{iN-N+1},\ldots\omega_{iN-1},\omega_{iN}).$$

Denote the corresponding shift by $\theta_N$, i.e.

$$(\theta_N\omega^N)_i = \omega_{i+1}^N .$$

The measure $Q^{N,0}$ induces a measure $\Psi_N(Q^{N,0})$ on $\Omega(M^N)$, which belongs to $M_\theta(\Omega(M^N))$, and has the property that the coordinates $\ldots\omega^N_{-1}$, $\omega^N_0$, $\omega^N_1,\ldots$ form a sequence of independent random variables. Let us note a consequence of this fact. Consider

$$V_n = \frac{1}{n}\sum_{k=0}^{n-1} \delta_{\theta_N^{-k}}$$

as a measure on $(\Omega(M^N))^-$. Now let

$$\Omega^N_0 = \{\omega\colon \lim_{n\to\infty} V_n = V, \text{ under } V\ldots, \omega^N_{-1}, \omega^N_0 \text{ are independent }\}.$$

By the ergodic theorem $V_n \to (\Psi_N(Q^{N,0}))_{-\infty,0}$ $Q^{N,0}$ - a.e., and hence $Q^{N,0}(\Omega^{N,0}) = 1$. Setting $\Omega^{N,k} = \theta^k\Omega^{N,0}$ one obtains $Q^{N,k}(\Omega^{N,k}) = 1$ for every integer $k$. By the Kolmogorov zero-one law $Q^{N,0}(\Omega^{N,j})$ equals zero or one for each $j$, since $\Omega^{N,j} \in F_{-\infty,-n}$, $n = 1, 2,\ldots$. If $N_0$ is the least positive $j$ such that $Q^{N,0}(\Omega^{N,j}) = 1$, then $Q^{N,0}(\Omega^{N,j})$ as a function of $j$ has period $N_0$. The typical situation is $N_0 = N$, but $N_0 < N$ is possible, e.g. if under $Q$ the coordinates $\ldots \omega_{-1}, \omega_0, \omega_1\ldots$ are independent $N_0 = 1$. So the restriction of $Q^N$ to $\Omega^{N,k}$ is simply $c_N Q^{N,k}$, where $c_N$ equals $N/N_0$. Since $\Omega^{N,k} \in F_{-\infty,0}$

$$Q^N[A|F_{-\infty,0}] = Q^{N,k}[A|F_{-\infty,0}] \text{ on } \Omega^{N,k}, \quad A \in F, \tag{5.2}$$

and therefore,

$$Q^N[A|F_{-\infty,0}] = Q^{N,k}[A|F_{-N+k+1,0}] \tag{5.3}$$

$$= Q[A|F_{-N+k+1,0}] \text{ on } \Omega^{N,k}, \quad A \in F_{-N+k+1,k}, \quad k=0,1,\ldots,N-1 .$$

The first equality in (5.3) follows from (5.2) because by definition the $\sigma$-field $F_{-\infty,-N+k}$ is independent of $F_{-N+k+1,\infty}$ under $Q^{N,k}$, so that the last member of (5.2) agrees with the middle member of (5.3) for $A \in F_{-N+k+1,\infty}$. Since $Q^{N,k}$ agrees with $Q$ on $F_{-N+1+k,k}$ the last equality in (5.3) also follows. We will want to apply (5.3) for $A \in F_1$, but this case is not included in (5.3) when $k = 0$. For that case we have

(5.4)
$$Q^N[A|F_{-\infty,0}] = Q_1(A) \text{ on } \Omega^{N,0}, \ A \in F_1 .$$

The notations $Q_1^N(\cdot\,|F_{-\infty,0})$ and $P_1(\cdot\,|F_{-\infty,0})$ will denote the restrictions of $Q^N(\cdot\,|F_{-\infty,0})$ and $P(\cdot\,|F_{-\infty,0})$ respectively to $F_1$.

Let us set $F'_{-N+k+1}$ equal to the trivial $\sigma$-field $\{\Omega,\phi\}$ if $k = 0$, and $F'_{-N+k+1} = F_{-N+k+1,0}$, $k=1,2,\dots N-1$. Then we will obtain from (5.3) and (5.4) that

(5.5)
$$Q_1^N(\ |F_{-\infty,0}) = Q_1(\ |F'_{-N+k+1}) \text{ on } \Omega^{N,k}, \ k=0,1,\dots N-1 .$$

Note that

$$\infty > H_{P^*}(Q) = \int\int (\log \frac{d(Q_\omega)_1}{d(P_\omega)_1}) \quad Q_\omega(dy)Q(d\omega)$$

and Condition 2.5 ensures that

$$\infty > \int[\int \log \frac{d(Q_\omega)_1}{dP_1} Q_\omega(dy)]Q(d\omega) .$$

Since $Q_1 = \int (Q_\omega)_1 Q(d\omega)$, the fact that $h_\mu(\nu)$ (defined in (2.3)) is a convex function of $\nu$ implies

$$(5.6) \qquad \int \log \frac{dQ_1}{dP_1}\, dQ_1 < \infty \quad .$$

Note

$$H_{p^*}(Q^N) = \int_\Omega \{ \int_M \log \frac{dQ_1^N(\cdot|F_{-\infty,0})}{dP_1(\cdot|F_{-\infty,0})} Q_1^N(dx|F_{-\infty,0}) \} \, dQ^N$$

$$= \frac{1}{N} \sum_{k=0}^{N-1} \int_\Omega \{ \int_M \log \frac{dQ_1(\cdot|F'_{-N+1+k})}{dP_1(\cdot|F_{-\infty,0})} Q_1(dx|F'_{-N+1+k}) \} \, dQ^{N,k}$$

$$= \frac{1}{N} \sum_{k=0}^{N-1} \int_\Omega \{ \int_M \log \frac{dQ_1(\cdot|F'_{-N+1+k})}{dP_1(\cdot|F'_{-N+1+k})} Q_1(dx|F'_{-N+1+k}) \} \, dQ^{N,k}$$

$$+ \frac{1}{N} \sum_{k=0}^{N-1} \int_\Omega \{ \int_M \log \frac{dP_1(\cdot|F'_{-N+1+k,0})}{dP_1(\cdot|F_{-\infty,0})} Q_1(dx|F'_{-N+1+k}) \} \, dQ^{N,k}$$

$$\equiv \sum\nolimits_1^{(N)} + \sum\nolimits_2^{(N)} \quad .$$

In the expression defining $\sum_1^{(N)}$ the expression within braces is $F'_{-N+1+k}$- measureable. On this $\sigma$-field $Q^{N,k}$ coincides with $Q$, so that

$$\sum\nolimits_1^{(N)} = \frac{1}{N} \sum_{k=0}^{N-1} \int_\Omega \{ \int_M \log \frac{dQ_1(\cdot|F'_{-N+k+1})}{dP_1(\cdot|F'_{-N+k+1})} Q_1(dx|F'_{-N+k+1}) \} dQ .$$

For fixed $N$ denote the (k+1)th term in the sum by $H^{N,k}$ so that

$$\sum\nolimits_1^{(N)} = \frac{1}{N} \sum_{k=0}^{N-1} H^{N,k}$$

For $1 \le k \le N-1$, $H^{N,k}$ is an approximation to $H_{p^*}(Q)$; indeed $H^{N,k} = H_p^{N-k-1}(Q)$ in the notation of Theorem 4.2. Hence, $H^{N,k} = H^{N-k-1}$, $1 \le k \le N-1$, $N=1, 2,\ldots$ . As pointed out in Theorem 4.2, under Condition 4.1 (which is part of the condition 5.2 we are assuming) $H^n$ converges to $H_{p^*}(Q)$ as $n \to \infty$ . The term $H^{N,0}$ is a finite number, independent of N, according to (5.6). Thus $\sum_1^{(N)} = N^{-1}[H^{N,0} + H^{N-2} + H^{N-3} + \ldots H^0] \to H_{p^*}(Q)$

as $N \to \infty$.

To treat $\sum_2^{(N)}$ recall (4.2) and Remark 5.3 to write

$$\sum_2^{(N)} = \frac{1}{N} \sum_{k=0}^{N-1} \int_\Omega \{ \int - \sum_{j=N-k+1}^{\infty} \log(1+\alpha_j) Q_1(dx|F_{-N+k+1})\} dQ^{N,k}$$

$$+ \frac{1}{N} \int_\Omega \{ \int [ - \sum_{j=0}^{\infty} \log(1+\alpha_j) - \log(1+\alpha)] Q_1(dx)\} dQ^{N,0} .$$

It follows from Condition 5.2 (even Condition 4.2) that $\sum_{j=n}^{\infty} \log(1+\alpha_j(x,\omega))$ converges to zero uniformly in $x$ and $\omega$ as $n \to \infty$. Condition (5.2) also ensures that $|\log(1+\alpha_j(x,\omega))|$ and $|\log(1+\alpha(x,\omega))|$ are bounded, uniformly in $x$, $\omega$ and $j$. Consequently $\sum_2^{(N)}$ approaches zero as $N \to \infty$.

5.6 THEOREM. *Let* $A = (\Omega, F, P, \theta)$ *be a shift with* $P \in M_\theta'$ *satisfying Condition* 5.2 *and associated* $P^*$ *chosen as in Theorem* 4.2 . *Then* $H_{P^*}$ *is an upper deviation function.*

PROOF. We need to verify (5.1) with $H_{P^*}$ for K. Assume $H_{P^*}(Q) < \infty$, as otherwise there is nothing to prove. Then $Q \ll_{loc} P$ by Theorem 5.5. If $Q \in M_{\theta,e}$ the conclusion follows from Proposition 5.1. If $Q \in M_\theta$ let $Q^N$ be as in Theorem 5.5. Since $N$ in (5.1) is a neighborhood of Q, one also has $Q^N \in N$ for N big enough. Now Proposition 5.1 applies to $Q^N$, and letting $N \to \infty$ and using Theorem 5.5 completes the proof.

5.7. COROLLARY. *If the condition of Theorem* 5.6 *hold and* $L_n'$ *satisfies the compactness condition* 3.2 , *then* $\hat{H}_{P^*}$ *is the deviation function for* S.

PROOF. The corollary follows at once from Theorem 5.6, Theorem 3.3, and (vii) of Proposition 1.1. □

We now discuss another approach for obtaining the lower bound, following ideas of Donsker and Varadhan [DV, part 4]. Let $P \in M'_\theta$ with associated regular conditional probabilities $(P^*_\omega)$. Suppose $Q \in M_{\theta,e}$ with $H_{P^*}(Q) < \infty$. Then, by (iii) of Theorem 2.1, $(Q_{\bar{\omega}})_{0,n} \ll (P_{\bar{\omega}})_{0,n}$ for $Q$ - a.e. $\bar{\omega}$, $n=1,2,\ldots$, and we denote the corresponding Radon-Nikodym deviative by $\Psi_n(\omega)$; (hence $\Psi_n(\omega) = \Psi_n(\bar{\omega};\omega_1,\ldots,\omega_n)$). Then

$$\Psi_{n+m} = \Psi_n \cdot \Psi_m \circ \theta^n$$

so that the erogodic theorem immediately implies the following "conditional Shannon-McMillan Theorem",

$$\lim_{n\to\infty} \frac{1}{n} \log \Psi_n \to H_{P^*}(Q) \text{ , } Q \text{ - a.e.} \tag{5.7}$$

Now

$$P^*_{\bar{\omega}}[L'_n \in N] \geq \int_{L'_n \in N} e^{-\log \Psi_n} dQ_{\bar{\omega}} \geq$$

$$e^{-n(H_{P^*}(Q)+\varepsilon)} Q_{\bar{\omega}}[L'_n \in N \text{ , } \frac{1}{n}\log \Psi_n < H_{P^*}(Q) + \varepsilon] .$$

Let $N$ be a neighborhood of $Q$; then as $n$ approaches infinity the second factor in the last member tends to 1 $Q$ - a.e., so that

$$\underline{\lim}_{n\to\infty} \frac{1}{n} \log P^*_{\bar{\omega}}[L'_n \in N] \geq -H_{P^*}(Q) \text{ , } Q \text{ - a.e.} \tag{5.8}$$

Recall that we would like to obtain (5.1) with $H_P{}^*$ in place of K. Since (5.8) holds only Q - a.e. and typically P and Q are singular, we have a long way to go. The following condition will help.

5.8 CONDITION. *There exists a function* m *, from the positive integers to the positive integers, such that*

$$\lim_{n\to\infty} \frac{m(n)}{n} = 0$$

*and*

$$\lim_{n\to\infty} \frac{1}{n} \sup\{\,|\log \frac{P^*_{\eta^-}(A)}{P_{\omega^-}(A)}| \;:\; \omega^- \in \Omega^-,\ \eta^- \in \Omega^-,\ A \in F_{m(n),n}\} = 0 .$$

5.9 PROPOSITION. *Assume* $P \in M'_\theta$ *with associated* $P^*$ *satisfying condition* 5.8. *Let* $Q \in M_{\theta,e}$, *and let* $N$ *be a neighborhood of* Q . *Then the* $\underline{\lim}$ *assertion in* (5.8) *holds uniformly for* $\omega^- \in \Omega^-$. *In particular* (5.1) *holds with* $H_P^*$ *in place of* K.

PROOF. There exist neighborhoods $N'$ and $N''$ of Q , and a positive integer n' such that $N \supseteq N' \supseteq N''$ and

$$(5.9) \qquad [L'_n \in N] \supseteq \theta^{-m}[L'_n \in N'] \supseteq [L'_n \in N''],\ n \geq n'.$$

According to (5.8) there exists an $\omega^- \in \Omega$ and for every $\varepsilon > 0$ a number $n_\varepsilon$ such that

$$(5.10) \qquad \frac{1}{n} \log P^*_{\omega^-}[L'_n \in N''] \geq [H_{P*}(Q) + \frac{\varepsilon}{2}],\ n \geq n_\varepsilon$$

Now it follows from (5.9) and Condition 5.8 that there exists $n'_\varepsilon$ such that

$$\frac{1}{n}\log P^*_{\eta^-}[L'_n \in N] \geq -[H_{P^*}(Q) + \varepsilon] , \; n \geq n'_\varepsilon , \quad \eta^- \in \Omega^- . \quad \square$$

5.10 THEOREM. *Let* $S=(\Omega,F,P,\theta)$ *be a shift,* $P \in M'_\theta$ *, and associated* $P^*$ *satisfying Condition* 5.8 . *Let* $Q \in M_\theta$ *, and let* $N$ *be a neighborhood of* $Q$ . *Then the* <u>lim</u> *assertion in* (5.5) *holds uniformly for* $\omega^- \in \Omega^-$ . *In particular* (5.1) *holds with* $H_{P^*}$ *in place of* $K$ *, so that* $H_{P^*}$ *is an upper deviation function.*

PROOF. By (iv) of Theorem 2.1 we may assume that $Q$ is a finite convex combination of members of $M_{\theta,e}$ . We illustrate the argument for $Q = \frac{1}{2}(Q_1 + Q_2)$ , $Q_1 \in M_{\theta,e}$, $Q_2 \in M_{\theta,e}$ . Choose neighborhoods $N_1$ of $Q_1$ and $N_2$ of $Q_2$ such that $\frac{1}{2}(N_1 + N_2) \subseteq N$. For $\varepsilon>0$ , Proposition 5.9 allows us to choose $N_\varepsilon$ such that

$$\frac{2}{n}\log P^*_{\omega^-}[L'_{n/2} \in N_i] \geq -(H_{P^*}(Q_i) + \varepsilon ), \; i=1,2 , \; n \geq n_\varepsilon , \quad \omega^- \in \Omega^- .$$

Now use

$$P^*_{\omega^-}[L'_n \in N] \geq P^*_{\omega^-}[L'_{n/2} \in N_1] \cdot \inf_{\eta^-} P^*_{\eta^-} [L'_{n/2} \in N_2] . \quad \square$$

5.11 THEOREM. *Let* $S = (\Omega,F,P,\theta)$ *be a shift,* $P \in M'_\theta$ *and assume either Condition* 5.8 *or Condition* 5.2 *and also* (3.2) *and* (3.3) . *Then* $H_{P^*}$ *is the deviation function for* $S$ . *Furthermore, under Condition* 5.8, *the upper and lower bounds continue to hold if* $P[L_n \in A]$ *is* *replaced by* $P_{\omega^-}[L_n \in A]$, *and indeed the bounds hold uniformly as* $\omega$*-ranges over any compact set.*

PROOF. The first assertion is immediate from Theorem 5.6, Theorem 5.10, and Theorem 3.3. For the uniformity assertion use the uniformity in Theorem 5.10, the uniformity in Lemma 3.2, and an argument like that in Proposition 3.1 to go from $L$ to $L'$.

## 6. Examples

The scope of our results will be illustrated by a number of examples.

### 6.1 Example: Discrete shift.

Consider a shift $S = (\Omega, F, \theta, P)$ with $\Omega = \Omega(M)$ where $M$ is a finite or countable set endowed with the discrete topology, and $P \in M_\theta'$. Let

$$P_n^*(i_1 | i_{-n} \ldots i_{-1}, i_0) = P[X_1 = i_1 | X_{-n} = i_{-n}, \ldots X_{-1} = i_{-1} X_0 = i_0]$$

Condition 5.1 amounts to the following: for some positive constant $c_0$,

$$\lim_{n\to\infty} P_n^*(\omega_1 | \omega_{-n}, \ldots \omega_{-1}, \omega_0) = P^*(\omega_1 | \ldots \omega_{-1}\omega_0) \geq c_0 P_0(\omega_1) \tag{6.1}$$

and the convergence in (6.1) is uniform with respect to $\omega$. Then (3.3) holds, and so does the condition of Theorem 3 4 and according to Theorem 5 11 $H_{p*}$ is the deviation function.

It is interesting that there are shifts on a 2-element space not possessing a deviation function. Professor A. Sokal has shown me an example (based on modifying the examples of Kieffer [K]) in which the limit $C(Y)$ introduced in Corollary 1.7 fails to exist for $Y$ the coordinate function $(Y(\omega) = \omega_1)$.

## 6.2 Example. Piecewise monotone transformation of [0,1].

Now $S = ([0,1],\beta,T,\mu)$ where $\beta$ is the class of Borel sets on [0,1], $Tx = \phi(x)$ where $\phi$ is a piecewise monotone map of [0,1] into itself. We shall use the results of Adler [A] and some extensions given in [CFS]. Specifically assume that [0,1] is a union of a finite or countable number of pairwise disjoint intervals $\Delta_1, \Delta_2, \ldots$ such that on each $\Delta_i$ the function is strictly monotone ( $\phi$ may be increasing on some $\Delta_i$ and decreasing on others). Assume that $\phi$ is continuous on each $\Delta_i$ and has continuous second derivatives on the interior of each $\Delta_i$. Let $T^n x = \phi^{(n)}(x)$ be the nth iterate of $\phi$, and write $\phi'(x)$, $\phi''(x)$ for the first and second derivative of $\phi$ with respect to $x$. Assume that $\phi$ satisfies

(i) $$T(\Delta_i) = (0,1) , \; i = 1,2,\ldots$$

(ii) $$\text{For some } s, \inf_{\Delta_i} \inf_{x \in \Delta_i} (\phi^{(s)})'(x) = \lambda > 1$$

(iii) $$\sup_i \sup_{x_1,x_2 \in \Delta_i} \frac{\phi''(x_1)}{\phi''(x_2)} = C < \infty$$

Then it is known that there exists $\mu \in M_T$ such that $\mu$ and Lebesgue measure $\rho$ are mutually absolutely continuous and for some positive constant $c_o$

(6.2) $$\frac{1}{c_o} \leq \frac{d\mu}{d\beta} \leq c_o$$

and with respect to this invariant measure T is an exact endomorphism, Theorem 4, [CFS], p. 290. Hence $\mu \in M_{T,e}$ , and it follows that $\mu$ is uniquely determined, and our system $S = (\Omega, F, T, \mu)$ is specified.

Let us denote

$$\Delta_{i_1 i_2 \cdots i_n} = \Delta_{i_1} \cap T^{-1} \Delta_{i_2} \cap \ldots \cap T^{-n+1} \Delta_{i_n}$$

One has the following inequalities

(6.3) $$c_o^{-4} \mu(A) \leq \mu(T^{-n}A | \Delta_{i_1 i_2 \cdots i_n}) \leq c_o^4 \mu(A)$$

valid for any $A \in F$, and any $i_1, i_2, \ldots i_n \in M$, $n=1,2,\ldots$ . The first inequality follows from (6), p. 291 of [CFS] together with (6.2); the second inequality is proved in the same way.

Now define a stationary one-sided shift

$$S_1 = (\Omega^+ (M), F^+, \theta, P)$$

where $M = \{0,1,\ldots,r\}$ if there are $r$ intervals $\Delta_i$ , and $M = \{0,1,\ldots\}$ if there are denumerably many $\Delta_i$, and $P \in M_\theta$ is determined by requiring

$$P[\omega_1=i_1,\omega_2=i_2,\dots\omega_n=i_n] = \mu(\Delta_{i_1 i_2 \cdots i_n}).$$

Put

$$\Delta_\omega \equiv \Delta_{\omega_1\omega_2\cdots} \equiv \lim_{n\to\infty} \Delta_{\omega_1\omega_2\cdots\omega_n}$$

Using (6.3) one deduces that $\Delta_\omega$ cannot have positive length, so $\Delta_\omega$ contains at most one point, which will also be denoted by $\Delta_\omega$. This suggests the homomorphism of $S_1$ onto $S$ given by $\Psi(\omega) = \Delta_\omega$. Note that $x \in \Delta_\omega$ implies $Tx \in \Delta_{\theta\omega}$. As in the representation of reals by decimal expansions, our mapping $\Psi$ is almost a 1-1 map of sequence space onto [0,1]. However in certain cases $\Delta_\omega$ may be empty (so $\phi(\omega)$ is undefined); this will happen only if $\omega$ "corresponds" to an end-point of an interval $\Delta_{i_1 i_2 \cdots i_n}$, so for only denumerably many $\omega$. Call this set of bad $\omega$ $\Omega_0^+$, and note $\theta^{-1}\Omega_0^+ \subsetneq \Omega_0^+$. We would like to use $\Omega_0^+$ as our negligible set in the definition of homomorphism (Section 1). For this it suffices that $Q(\Omega_0^+) = 0$ for $Q \in M_\theta$. Suppose otherwise. Then $Q(\{\omega\}) > 0$ for some $\omega \in \Omega_0$ and since $Q \in M_\theta$, $Q(\{\omega\}) \leq Q(\{\theta\omega\}) \leq Q(\{\theta^2\omega\}) \leq \dots$ and this leads to a contradiction unless $\omega$ is periodic. Observe now that $\omega \in \Omega_0^+$ is not periodic (to see this one may examine the special case discussed in Example 6.2.1).

Since we have a homomorphism from $S_1$ to $S$, it remains only to show that $S_1$ has a deviation function. Although $S_1$ is a one-sided shift we can of course extend it to a two sided shift $S_2$ on $\Omega(M)$. $S_2$ is a discrete shift as discussed in Example 6.1, but I do not know if the convergence condition (6.1) necessarily holds. What (6.3) tells us is that there is a way of defining $P[\ |F_{-n,0}]$ such that

$$c_0^{-4}\, P(A) \le P[A|F_{-n,0}](\omega) \le c_0^{4}\, P(A), \quad \omega \in \Omega(M),\ A \in F_{1,\infty}$$

and this allows one to define $P^*$ so that the Condition 5.8 holds. When $M$ is infinite, one can verify the condition of Theorem 5.4. So, to show that $H_{P*}$ is also a lower deviation function we require (3.3). We do not know whether this is obtainable without further assumptions.

We now proceed to some specific examples where (3.3) does hold, so that $H_{P*}$ will be the deviation function.

### 6.2.1. Example: Binary Expansions:

Take $\phi(x) = 2x \bmod 1$, or more precisely

$$\phi(x) = \begin{cases} 2x, & 0 \le x \le 1/2 \\ 2x-1, & 1/2 \le x \le 1 \end{cases}$$

Here the coding process is of course the Bernoulli shift on two symbols: $M = \{0,1\}$ and $\omega_1, \omega_2, \ldots$ is a sequence of independent, identically distributed random variables, $P[\omega_i=1] = P[\omega_i=0] = 1/2$. The homomorphism is given by $\Psi(\omega) = \sum_{i=1}^{\infty} \omega_i 2^i$ and the negligible set of sequences are those ending in repeated 1, except that the sequences consisting of all 1's has been retained. The coding process, in this case the Bernoulli shift, is of course covered by the work of [DV, part 4]. so in this example much of the discussion given in Example 6.2 can be by-passed.

Consider the identity function $Id(x)$ on $[0,1]$. We know by Corollary 1.5 that the deviation function $k_{Id}$ exists; (recall that this controls the exponential rate of deviation for the partial sums $n^{-1} \sum_{k=0}^{n-1} T^k x$). In the present case one can obtain it explicitly using the homomorphism $\Psi$. That is we seek the deviation function for the partial sums $S_n = \sum_{r=0}^{n-1} \Psi \circ \theta^k$. Now a simple calculation shows that if $S'_n = \sum_{k=1}^{n} \omega_k$ then $|S_n - S'_n| < 2$, so that the desired deviation function agrees with that for $(S'_n)$; see (i) of Remark 1.8.

6.2.2. Example: Tent Map.

Now define $\phi$ by

$$\phi(x) = \begin{cases} 2x, & 0 \le x \le 1/2 \\ -2x + 2, & 1/2 \le x \le 1 \end{cases}$$

As in the preceding example the resulting system S is a homomorphic image of the Bernoulli shift on $\{0,1\}$ with equal probabilities $(1/2,1/2)$, but now the mapping is given by $\Psi(\omega) = \sum_{i=1}^{\infty} \hat{\omega}_i 2^i$ where $\hat{\omega}_1 = \omega_1$ and $\hat{\omega}_n = \omega_n$ if $\omega_1 + \ldots \omega_{n-1}$ is even, otherwise $\hat{\omega}_n = 1 - \omega_n$, $n=2,3,\ldots$. Calculating $k_{Id}$ explicitly appear difficult; however Griffeath [G] has shown me an ingenious approach which gives an expression for $k_{Id}$ amenable to computation. We can use Corollary 1.5 to gain some information about $k_{Id}$. If is a convex function vanishing at $x=1/2$, finite valued on $[0,2/3]$, and infinite on $(2/3,1]$. The vanishing at 1/2 follows from the fact that $\mu$ is Lebesgue measure, $\int x \, d\mu = 1/2$ and of course $K(\mu)=0$. Next note that the points 0 and 2/3 are fixed points, so the point masses $\delta_0$ and $\delta_{2/3}$ are members of $M_T$ with means 0 and 2/3 respectively, and they are in fact the only such measures, so that we need only calculate $K(\delta_0)$ and $K(\delta_{2/3})$, and on sequence space $\delta_0$ corresponds to $Q^0$= (point mass at 0,0...) while $\delta_{2/3}$ corresponds to $Q'$= (point mass at 1,1,...) so one easily finds

$$H_P*(Q^0) = H_P*(Q') = \log 2$$

so that $k_{Id}(0) = k_{Id}(2/3) = \log 2$. Since, $x > 2/3$ implies $(x + Tx)/2 < 2/3$ one sees that for $\gamma > 2/3$, $n^{-1}[x + Tx + \dots T^{n-1}x] < \gamma$ for all $x$ if $n$ is sufficiently big, so $k_{Id}(\gamma) = \infty$.

### 6.2.3. Example: Continued Fractions.

Let $\phi(x)$ equal the greatest integer in $x^{-1}$, $0 < x < 1$, $\phi(0)=0$. Then $Tx= \phi(x)$ is the transformation related to the continued fraction expansion of real numbers, (see [B] for more information). In this case the "coding process" is a shift on $\Omega^+(M)$, $M= \{1,2,\dots\}$ and $\omega$ gets mapped into $x$,

$$x = \cfrac{1}{\omega_1 + \cfrac{1}{\omega_2 + \dots}} .$$

This mapping is 1-1 and onto $[0,1]^*$ the set of all irrationals in $[0,1]$. We shall therefore consider our system $S$ as obtained by restricting $T$ to $[0,1]^*$, and using the relative topology there. That (3.3) holds follows from known explicit expressions, given by P. Levy, Bull. Soc. Math. 57, 1929, 178-194. It follows from the general case discussed in Example 6.2 that a deviation function $K$ exists. In this case $k_{Id}$ is finite exactly on $(0,(-1 + \sqrt{5})/2)$. Here $(-1 + \sqrt{5})/2$ corresponds to $\omega = (1,1,\dots)$ and it is the largest fixed point of $T$.

### 6.3. Example: Other shifts.

"Other" refers to shifts which do not satisfy the hypotheses for our lower bound results from Section 5. Nevertheless one may be able to obtain deviation functions by coding, i.e. by showing that the shift in question is the homomorphic image of a nice shift.

To illustrate, let us start with the Bernoulli shift on $M = \{0,1\}$ again, $P[\omega_1=0] = P[\omega_1=1] = 1/2$. Let $(c_n)$, $n=0,\pm1,\ldots$ be a sequence of reals with $\sum c_n \leq 1$. One can construct a new shift on $[0,1]$ by taking "moving averages", i.e. let $\hat{\omega} = (\hat{\omega}_n)$ be defined by

$$\hat{\omega}_n = \sum_{k=-\infty}^{\infty} c_k \omega_{n+k}$$

Putting $\phi(\omega) = \hat{\omega}$ and $\hat{P} = \phi(P)$, our new process is a shift $(\Omega([0,1]),\hat{P})$, and it is a homomorphic image of the Bernoulli shift. Consider some special cases:

(a) $c_n = 2^{-n}$, $n=1,2\ldots$, $c_n = 0$ for $n \leq 0$. This is essentially the same as example 6.2.1 Here however we actually have a shift. The shift is deterministic ( $\hat{\omega}_1$ is determined by $\hat{\omega}_0$). Since we are in a Markovian situation there is an unambiguous definition of the extropy function $H_P(\cdot)$, but this is not the deviation function in this case.

(b) $c_n = 2^{-n-1}$ for $n \leq 0$; $c_n = 0$ for $n > 0$. Our shift $(\hat{\omega}_n)$ is a Markov chain on $[0,1]$ which moves from $x$ to $x/2$ with probability 1/2 and to $1/2 + x/2$ with probability 1/2. In this case one also obtains that $k_{Id}$ coincides with the deviation function for $(S_n')$, where $S_n' = \omega_1 + \ldots \omega_n$; the reason is the same as in Example 6.2.1.

Markov chains similar to the one encountered in (b) have been studied. That is chains on $[0,1]$ for which the process moves from $x$ to one of $r$ new positions ($r$ fixed), $\rho_1(x),\ldots\rho_r(x)$ with respective probabilities $p_1(x),\ldots p_r(x)$ . It would be very interesting to find a method of coding some class of chains of this type by nice shifts.

(c) $c_n = 2^{-n-2}$, $n \leq 0$, $c_n = 2^{-n-1}$, $n \geq 1$. In contrast to the other examples we have dealt with the homomorphism here is genuinely many to one.

## 6.4. Example: Non-convex K.

Consider the shift $S=(\Omega(M), F, \theta, P)$ with $M=\{-1,1\}$ and $P(\{\ldots 1,1,1,\ldots\}) = P(\{\ldots 0,0,0,\ldots\}) = \frac{1}{2}$. In this case it is obvious that the deviation function is given by $K(Q) = 0$ if $Q$ is the point mass at $(\ldots 1,1,1,\ldots)$ or if $Q$ is the point mass of $(\ldots 0,0,0\ldots)$ and $K(Q) = \infty$ otherwise. Note that K is not convex. (This example is essentially the same as one used by Ellis [E] to illustrate a related point).

## Bibliography


[A] R. ADLER. F-expansions revisited. Lecture Notes in Math. 318, pp. 1-5. Springer-Verlag, New York - Berlin - Heidelberg, 1973.

[Az] R. AZENCOTT. Grandes deviations et applications. Lecture Notes in Math. 774, pp. 1-249. Springer-Verlag, New York - Berlin - Heidelberg, 1980.

[BGZ] R.R. BAHADUR, J.C. GUPTA, S.L. ZABELL. Large deviations, tests and estimates, Asymptotic theory of statistical tests and estimation, (ed. I.M. Chakravarti). Academic Press, New York, 1979, 33-64.

[BZ] R.R. BAHADUR, S.L. ZABELL. Large deviations of the sample mean in general vector spaces. *Ann. Prob.* 7 (1979) 587-621.

[BP] V. BARBU, Th. PRECUPANU. *Ccnvexity and Optimisation in Banach Spaces*. Editura Academiei, Bucharest, 1978.

[B] P. BILLINGSLEY. *Ergodic Theory and Information*. Wiley, New York - London - Sidney, 1965.

[C] H. CRAMER. Sur un nouveau theoreme limite de la theorie des probabilites. Colloquium on theory of probability. Paris-Hermann (1937).

[CFS] I.P. CORNFELD, S.V. FOMIN, Ya.G. SINAI. *Ergodic Theory*. Springer-Verlag, New York - Berlin - Heidelberg, 1982.

[CG] J.T. COX, D. GRIFFEATH. Large deviations for Poisson systems of independent random walks. Preprint.

[DGS] M. DENKER, C. GRILLENBERGER, K. SIGMUND. Ergodic theory on compact spaces. Lecture Notes in Math 527, New York - Berlin - Heidelberg, 1976.

[DV] M. DONSKER, S.R.S. VARADHAN. Asymptotic evaluation of certain Markov process expectations for large time. *Comm. Pure Appl. Math.* : part 1, 28 (1975), 1-47; part 2, 29 (1976), 279-301; part 3, 29 (1976), 389-461; part 4, 36 (1983), 183-212.

[E] R. ELLIS. Large deviations for a general class of random vectors. *Ann. Prob.* 12(1984), 1-11.

[G] D. GRIFFEATH. Private communication.

[K] J.C. KIEFFER. A counterexample to Perez's generalization of the Shannon-McMillan Theorem. *Ann. Prob. 1* (1973) 362-364; *Ann. Prob. 4* (1976), 153-154.

[M] SHU-TEH C. MOY. Generalisations of Shannon-McMillan Theorem. *Pacific J. Math.* (1961), 705-714.

[P] A. PEREZ. Extensions of Shannon-McMillan Theorem to more general stochastic processes. *Trans. Third Prague Conference on Information theory*, Czech. Acad. Sci., Prague 1964, 545-575.

[$P_1$] A. PEREZ. McMillan's limit theorem for pairs of stationary random processes. *Kybernetica, 16* (1980) 301-314.

[T] Y. TAKAHASHI. Entropy function (free energy) for dynamical systems and their random perturbations. *Proc. Int. Symp. SDE*, Kyoto 1982, to appear.

[V] S.R.S. VARADHAN. Large deviation and applications. Preprint, to appear in SIAM CBMC-NSF Regional Conference in Applied Math series.

S. OREY
Department of Mathematics
University of Minnesota
Minneapolis, Minnesota 55455

## Already Published in Progress in Probability and Statistics

PPS 1 **Seminar on Stochastic Processes, 1981**
E. Çınlar, K. L. Chung, R. K. Getoor, editors
ISBN 3-7643-3072-4, 248 pages, hardcover

PPS 2 **Percolation Theory for Mathematicians**
Harry Kesten
ISBN 3-7643-3107-0, 432 pages, hardcover

PPS 3 **Branching Processes**
S. Asmussen, H. Hering
ISBN 3-7643-3122-4, 472 pages, hardcover

PPS 4 **Introduction to Stochastic Integration**
K. L. Chung, R. J. Williams
ISBN 0-8176-3117-8
ISBN 3-7643-3117-8, 204 pages, hardcover

PPS 5 **Seminar on Stochastic Processes, 1982**
E. Çınlar, K. L. Chung, R. K. Getoor, editors
ISBN 0-8176-3131-3
ISBN 3-7643-3131-3, 302 pages, hardcover

PPS 6 **Least Absolute Deviation**
Peter Bloomfield, William L. Steiger
ISBN 0-8176-3157-7
ISBN 3-7643-3157-7

PPS 7 **Seminar on Stochastic Processes, 1983**
E. Çınlar, K. L. Chung, R. K. Getoor, editors
ISBN 0-8176-3293-X
ISBN 3-7643-3293-X, 290 pages, hardcover

PPS 8 **Products of Random Matrices with Application to Schrödinger Operators**
Philippe Bougerol, Jean Lacroix
ISBN 0-8176-3324-3
ISBN 3-7643-3324-3, 296 pages, hardcover